Please return / renew by date shown.
You can renew at: **norlink.norfolk.gov.uk**
or by telephone: **0344 800 8006**
Please have your library card & PIN ready.

NORFOLK LIBRARY
AND INFORMATION SERVICE

BASIC CONCEPTS IN
MODERN
MATHEMATICS

JOHN EDWARD HAFSTROM

DOVER PUBLICATIONS, INC.
MINEOLA, NEW YORK

Bibliographical Note

This Dover edition, first published in 2013, is an unabridged republication of the work originally published by Addison-Wesley Publishing Company, Inc., Reading, Massachusetts, in 1961.

International Standard Book Number

ISBN-13: 978-0-486-49729-7
ISBN-10: 0-486-49729-1

Manufactured in the United States by Courier Corporation
49729101
www.doverpublications.com

To Regina

PREFACE

Since the first Sputnik there has been a significant increase in the number of colleges that have added a mathematics requirement to the undergraduate program for students majoring in the social studies or the humanities, students who have traditionally shunned college mathematics. A special one-quarter or one-semester course is ordinarily offered; a terminal course apart from the freshman year sequence undertaken by science majors. If one examines the descriptions of such terminal courses in various college bulletins, it is apparent that they are usually survey courses from which students gain superficial knowledge about many topics. Too often such courses are so designed that any but the most inept student can "get by"; too often the student undertaking such a terminal course meets the letter but not the spirit of his college's mathematics requirement. I believe that a more realistic terminal mathematics course, in keeping with the spirit of the times, is one that focuses on a relatively few fundamental concepts, but delves into each deeply enough to challenge the student; this book has been written for such a course. Of the many important topics which vie for inclusion in a book of this kind, I have selected a few which lend themselves to the construction of number systems. This method of selection seems natural since I believe it is advisable to include topics having a ready application.

The Mathematics Department of the University of Minnesota, Duluth offers a terminal course (*Modern Basic Mathematics*) which attempts to acquaint the student with a few fundamental mathematical ideas. This course, undertaken by undergraduates partially to satisfy general education requirements in the science area, carries five quarter credits, and is based on the material in this book. Omitting the axiom of induction in Chapter 2 and a detailed discussion of the longer demonstrations in Chapter 7, we are able to cover the first seven chapters quite thoroughly in fifty class meetings, eight of which are devoted to review and examination. Our experience with this course has been most gratifying: (1) we have found a growing preference for this course over a traditional intermediate algebra course which students may elect to satisfy their general education requirement; (2) we have found an occasional student having an alleged aversion to mathematics who rises above his aversion to meet the challenge of unfamiliar ideas and decides that mathematics can be interesting; (3) we have found an occasional student who discovers for himself that he has mathematical talent and elects to take further courses in mathematics. Most of the students undertaking the terminal course are freshmen having a background of one year of elementary algebra and one year of plane geometry; we have found that this background is adequate, and that students who put forth honest effort can make sense out of the material.

A further possible use for this book should be mentioned here. The material in Chapter 8 (The Rational Numbers) and Chapter 9 (The Real Numbers) is obviously not suitable for the terminal course described in the previous paragraph, but the inclusion of these chapters makes the book appropriate for an introductory course in modern mathematics designed for students who have completed a year of calculus. Such a course could serve as a bridge between the traditional freshman-sophomore mathematics sequence and a course in modern abstract algebra based on, say, Birkhoff and MacLane's *A Survey of Modern Algebra* or a similar text. It is well known that many students who have performed fairly well in their freshman and sophomore mathematics courses have faltered in modern algebra; for many the plunge from the traditional to the modern seems too abrupt. Although the Department of Mathematics of the University of Minnesota, Duluth offers no such transitional course, we have twice used the material of this book in mimeographed form as the text for a summer course held for high-school science and mathematics teachers under the auspices of the National Science Foundation. The participants had little or no prior acquaintance with so-called modern mathematics. The material was appropriate and was well received by the students.

I wish to extend thanks to my colleagues in the Department of Mathematics at the University of Minnesota, Duluth who, from their classroom experience with several versions of the manuscript, contributed many helpful ideas. I wish especially to thank Dr. Ervin K. Dorff for his advice and encouragement. I wish also to thank the editorial staff of the Addison-Wesley Publishing Co. for their continuing help.

J. E. H.

Department of Mathematics and Engineering
University of Minnesota, Duluth

CONTENTS

CHAPTER 1

INTRODUCTION

1-1 Our program of study. In studying this text, it will be our purpose to become acquainted with certain modern mathematical concepts that will help us better to understand and use mathematics. In a text such as this, which attempts to give only an elementary treatment of these concepts, we shall be able to consider only a few of their applications; in particular, we shall use these concepts to gain a stronger understanding of the *structure* of some familiar number systems and how these systems are related. Among the concepts that the reader will encounter for perhaps the first time are those of *set, mapping, relation, group,* and *isomorphism.* It is hoped that application of these concepts to the study of number systems will result in a new appreciation for the logic of mathematical arguments and an added respect for the powers of the human mind.

1-2 How numbers developed. From our studies in elementary grades and in high school, we are all familiar with a variety of such numbers as *whole numbers, fractions, negative numbers,* and so on. It is interesting to note that our experience in learning about numbers, i.e., progressing at a rate that maturity allows, roughly parallels that of the human race, except that our learning experience is compressed into a short span of a few years!

Leopold Kronecker, a celebrated German mathematician of the nineteenth century, is supposed to have said, "God created the natural numbers; all else is the work of man." By "natural numbers" Kronecker meant the numbers 0, 1, 2, 3, . . . , which we often refer to as the "whole" numbers, or the *counting numbers.* We may ask whether we should take Kronecker's statement literally, i.e., whether we should believe God handed down to man the symbols '0,' '1,' '2,' '3,' . . . together with instructions for using them. Of course not; Kronecker simply meant that God gave man a mind and an *instinctive consciousness* of natural numbers. It almost seems that man is born to count; we know, for example, that even a very small child is *conscious* of the number of persons in his immediate family and, if he owns three identical teddy bears, is *conscious* of the fact when one is missing. Man, himself, devised the *symbols* which he uses to represent the numbers of which he is conscious. We have an intuitive notion of the meaning of "number" and "number system," but we shall make no attempt at this time to formulate definitions of

1

these concepts; after studying *sets* and *mappings* we shall be able to
state a technical definition of "number" (Chapter 4).

We have implied that the symbols '0,' '1,' '2,' '3,' . . . are *not* the numbers of man's consciousness; strictly speaking, they are *symbols* (or *names*)
which represent or "stand for" these numbers. However, by habit, we
call these symbols *the natural numbers* and we shall continue to do so.
Over the centuries, man has used many other sets of symbols, and the
familiar symbols '0,' '1,' '2,' '3,' . . . are of fairly recent origin in western
culture. The Romans used the symbols 'I,' 'II,' 'III,' 'IV,' 'V,' 'VI,' . . .
(the Roman numerals), and earlier civilizations used other sets of symbols
for the natural numbers.* The early Romans and other Mediterranean
peoples used no symbol for the natural number we call 'zero,' perhaps
because they were not yet fully aware that zero actually is a natural
number. Note that we are able to write any natural number we wish
by using no more than ten different symbols: '0,' '1,' '2,' '3,' '4,' '5,' '6,'
'7,' '8,' '9.' The Roman system of notation had no such feature; the
Romans wrote 'V' for '5,' 'L' for '50,' 'D' for '500,' and so on. In short,
in order to write larger and larger natural numbers, the Romans were
forced to invent more and more symbols to represent them. Computations we consider routine were very awkward for the Romans and they
relied heavily on the abacus,† mankind's first digital computing machine.
After the Crusades, the Hindu-Arabic symbols '0,' '1,' '2,' '3,' . . . were
introduced into western Europe, and the relatively simple rules (algorithms) for computing with them made computing a common and
widely practiced art.

Today we use numbers other than only the natural numbers (numbers
which, according to Kronecker, are "the work of man") and we shall
now briefly discuss how such numbers as fractions, negative numbers,
etc., came into use and how they were blended with the natural numbers
to form what we call the *real number system*.

Fractions were invented because, when man began to build temples and
survey land, for example, he found that whatever standard unit for
measuring length he adopted, whether it was the length of his foot, the
length of his forearm, or the length of the king's beard, he encountered
lengths which did not contain the standard unit a *natural* number of
times. Thus, he began to divide the standard units on his measuring
devices into halves, thirds, quarters, and so on, and *fractions* were born.
Records tell us that the Babylonians and Egyptians were using fractions
as early as 3000 B.C.

---

* J. Houston Banks, *Elements of Mathematics*, New York, Allyn and Bacon,
1956 (Sec. 2.1–2.6).

† Robert L. Swain, *Understanding Arithmetic*, New York, Rinehart, 1957
(Ch. 1).

The story of *negative numbers* is different. The early algebraists, mainly the Arabs, found that certain equations had no solution, if by "solution" one meant a natural number or a fraction. For example, if the only numbers familiar to him were the natural numbers and fractions, an algebraist confronted with the task of solving the equation $3x + 12 = 0$ would be forced to say, "This equation has no solution; the statement $3x + 12 = 0$ can never be true if x is to be a natural number or a fraction." To remedy this state of affairs, some bold and imaginative algebraist invented negative numbers and described the rules that must govern their use if they were to blend with the existing numbers and form, with them, a larger system of numbers which we know today as the *rational* number system. Today, we ordinarily call the numbers

$$\ldots, -4, -3, -2, -1, 0, 1, 2, 3, 4, \ldots$$

the *integers* and define the *rational numbers* to be the collection of all numbers of the form p/q, where p and q are integers and $q \neq 0$ ($\neq$ means "is not equal to"). An integer is itself a very special kind of rational number, i.e., one that may be written in the form $p/1$, where p is an integer. Today we take for granted the ease with which, by means of elementary algebra, we are able to find a solution for any equation of the form $ax + b = c$ whenever b and c are rational numbers and a is a rational number other than 0; few of us are aware of the debt we owe to the anonymous inventor of negative numbers. Because the need for negative numbers was not, at first, a *practical* one, such as the need for fractions, for centuries negative numbers were looked upon as freak numbers that algebraists used as playthings. Later, however, as man's knowledge of the physical world . increased, negative numbers became necessary for describing physical relationships and, by the time of the Renaissance, they had become accepted members in the society of numbers.

The need for still another kind of number first arose from the study of geometry. In about 500 B.C. Greek geometers proved that, using a given unit of length, they could construct a line segment whose length did not contain the given unit a *rational* number of times. For example, if we construct a right triangle whose legs are each one inch in length, the hypotenuse, by the Theorem of Pythagoras, will have a length equal to $\sqrt{2}$ (the "square root" of 2) inches, where $\sqrt{2}$ is a "number" whose square is 2. The number $\sqrt{2}$ is called an *irrational* number; that is, it is not possible to express this number as the *ratio* of two integers. In other words, it is not possible to find a number of the form p/q, where p and q are integers and $q \neq 0$, whose square is 2. Algebraists were also forced to deal with irrational numbers in solving quadratic equations; an innocent-appearing equation such as $x^2 - 5 = 0$ has two solutions, $\sqrt{5}$

and $-\sqrt{5}$, both of which are irrational numbers. Fortunately, it was found that it was possible to blend irrational numbers with rational numbers so that, together, they formed a larger family of numbers which we today call the *real number system*. This system now meets most of man's everyday needs. (The reader interested in knowing more about the historical development of the real number system is urged to consult such books on the history of mathematics as Howard Eves, *An Introduction to the History of Mathematics*, Rinehart.)

1-3 The mathematician's view of the development of numbers. The previous discussion of the development of numbers may have given the impression that negative numbers, rational numbers, and irrational numbers have always existed but had remained hidden from view, waiting only to be discovered so that, along with the natural numbers, they could be fitted together like pieces of a jigsaw puzzle to form the real number system. This is not the mathematician's point of view; remember Kronecker's quotation: ".... all else is the work of man." Most mathematicians prefer to think of the development of the real number system as a process of orderly and logical growth which progressed from the natural numbers to the integers, from the integers to the rationals, and, finally, from the rationals to the reals; they look at the development of the real numbers as they *would like to have planned it.* They are like the "Monday morning quarterbacks" who replay last Saturday's football game; they say, "We would have done things differently." Almost a hundred years ago mathematicians began to be bothered by the fact that real numbers developed without their supervision, like plants in an untended garden. Fortunately, however, they were able to *prove* that the end product (the real numbers) of this unsupervised growth is good; in other words, they proved that there is a logical plan which, had it been followed, would have yielded the real number system as we know it today. We shall learn something of this plan in Chapters 7, 8, and 9.

1-4 A word to the reader. If you have had a course in plane geometry, you may recall that your study began with a few such undefined concepts as "point" and "line," and these, together with basic definitions and assumptions (axioms), were used to prove geometric properties of triangles, circles, polygons, and so on. It was not permissible to make a statement unless it could be defended by quoting appropriate definitions, axioms, or previously proved theorems. In this text we shall follow a somewhat similar procedure; we shall start with certain undefined concepts, a few definitions, and some assumptions about the arithmetical properties (structure) of the natural number system. We shall then use these to construct several other number systems (particularly the integers, the rationals, and the reals) and to study their structures. Above all, we shall stress the importance of making precise statements and being able to defend them.

The importance of the role that definitions will play in our study cannot be overemphasized. Each new concept will usually be introduced by the statement of one or more definitions, and each new definition should be immediately *memorized*. Of course, mere memorization of a definition does not guarantee that it will be understood or appreciated, but it is a first step toward understanding. After each new definition is presented, several clarifying examples will usually be given to aid in understanding it, and perhaps to indicate how it is to be applied. At this point it will be helpful to try to restate the definition in your own words in as many ways as possible but, in rephrasing, you must be careful to neither add nor subtract meaning from the definition as originally stated. If at first the significance of a new definition is difficult to grasp, you will find it rewarding to return to it again and again. If you thoroughly understand each step as it is presented, your work will be made easier in the long run.

1–5 Numbers and numerals. Doubtless, the reader has observed (Section 1–2) that single quotes were used to make a distinction between the *numbers* 0, 1, 2, 3, . . . and the *symbols* (or *names*) '0,' '1,' '2,' '3,' . . . for these numbers. Logically, this distinction is one that should always be made for there *is* a difference between a thing and its name. We wrote 0, 1, 2, 3, . . . when the reader was to think of the natural numbers and '0,' '1,' '2,' '3,' . . . when he was to interpret 0, 1, 2, 3, . . . as *symbols* or *names* for these numbers. The symbols '0,' '1,' '2,' '3,' . . . are called *numerals*.

In the chapters that follow we will *not* distinguish between things and their names and, in particular, we will not distinguish between numbers and numerals. In other words, we will omit single quotes when writing numerals. In cases where a distinction between *number* and *numeral* should be made it will be left to the reader to determine from the context whether or not single quotes logically ought to be present. For example, the reader should be able to correctly interpret 2 as the *number* 2 or the *numeral* '2' from the context in which it occurs. For further reading on the distinction between things and the names of things, the excellent book *An Introduction to Modern Mathematics* by Robert W. Sloan (Prentice-Hall, 1960) is recommended. In particular, refer to Chapter 1.

CHAPTER 2

THE NATURAL NUMBERS

2–1 Introduction. When we write the natural numbers (the counting numbers) as 0, 1, 2, 3, 4, 5, . . . , the three dots indicate that something has been omitted and that, if we wished, we could continue the list further. One thing is immediately apparent, however; the list of natural numbers can never be continued to the point of writing the *last* natural number, for there is no *last* natural number. We characterize this property of the list of natural numbers by saying that the list is *infinite*. We shall say more about the adjective "infinite" and its opposite, "finite," in our discussion of sets in Chapter 3.

We are all familiar with the addition ($+$) and multiplication ($\cdot$) of natural numbers; these are called "operations on the natural numbers." A precise definition of "operation" will be given in Chapter 4, but for now we will content ourselves with knowing how to add and how to multiply natural numbers, and with the knowledge that adding or multiplying any two natural numbers will always yield another natural number.

The relation $=$ ("is equal to") should be clearly understood and we state the following definition:

> DEFINITION. If a and b are symbols for natural numbers, then we write $a = b$ if, and only if, a and b are symbols for the *same* natural number. If a and b are symbols for *different* natural numbers, we write $a \neq b$.

Any natural number can be properly represented by any one of several symbols. For example, $3 + 4$, $5 + 2$, and $0 + 7$ all symbolize the same natural number as the symbol 7. Thus when we write $5 + 2 = 7$, we are using the equality symbol ($=$) in the same way that we use it when we write Chicago = The Windy City, or Minnesota = The Gopher State; in each case the symbol or name on the left "stands for" the *same* thing as the symbol or name to the right of the equality sign.

We have called $=$ ("is equal to") a "relation" on the natural numbers; in other words, two natural numbers may be "related" by "being equal." The concept of "relation" is a very important one, and we will discuss it in some detail in Chapter 6. For the present, however, we shall talk about $=$ being a "relation" on the natural numbers even though we lack a full understanding of the concept.

The relation $=$ on the natural numbers possesses the following three properties:

(1) *Reflexive property.* If a is any natural number, then $a = a$.

(2) *Symmetric property.* If a and b are any natural numbers such that $a = b$, then $b = a$.

(3) *Transitive property.* If a, b, c are any natural numbers such that $a = b$ and $b = c$, then $a = c$.

Because the relation $=$ has these three properties (reflexivity, symmetry, transitivity) we call it an "equivalence relation." We will encounter many examples of equivalence relations in our later work.

We recall from elementary algebra, where we first formally met the *substitution principle*, that "equals may be substituted for equals." This principle, although used in mathematics, is not only a mathematical principle, but a principle of logic that we use unconsciously in everyday conversation, writing, and thinking. For example, the newspaper reporter who writes "Record Snowfall Hits Gotham" has used the substitution principle in replacing the name "New York City" by the nickname "Gotham." We now state the substitution principle formally as it applies to natural numbers; it will also apply to other kinds of numbers when we meet them.

THE SUBSTITUTION PRINCIPLE. If a and b are symbols for the same natural number (that is, if $a = b$), then any true (or false) statement involving a will remain true (or false) if a, *anywhere* it appears in the statement, is replaced by b.

Consider the following examples, which illustrate how the substitution principle is used.

EXAMPLE 1. $7 + 5 = 12$ is a *true* statement. Since $8 + 4 = 12$, we may, by the substitution principle, replace 12 by $8 + 4$ to obtain the *true* statement $7 + 5 = 8 + 4$.

EXAMPLE 2. $8 = 5$ is a *false* statement. Since $5 = 2 + 3$, we may, by the substitution principle, replace 5 by $2 + 3$ to obtain the *false* statement $8 = 2 + 3$.

EXAMPLE 3. $10 + 7 \neq 10 + 4$ is a *true* statement. Since $8 + 2 = 10$, we may, by the substitution principle, replace 10, *anywhere* it appears in the statement, by $8 + 2$. We are completely free to replace one 10 or the other, or both, by $8 + 2$. We may, for example, replace only the 10 on the left by $8 + 2$ to obtain the *true* statement $(8 + 2) + 7 \neq 10 + 4$. (We have used parentheses to indicate that $8 + 2$ has replaced a *single* number.)

Observe that in each of the above examples we had a statement involving a natural number a [$a = 12$ in (1), $a = 5$ in (2), and $a = 10$ in (3)]; in each statement we replaced a by b [$b = 8 + 4$ in (1), $b = 2 + 3$ in (2), and $b = 8 + 2$ in (3)], and each statement retained its truth or falsity.

We should be reminded that the substitution principle is a principle of logic and its applications are not confined to the field of mathematics. However, we will be most interested in the substitution principle as it applies to statements about numbers.

2–2 Axioms. In Section 1–4 we mentioned that our study would begin with, among other things, some assumptions about the arithmetical properties of the natural number system. These assumptions we call *axioms* and *we shall accept them without proof.* By using these axioms we shall be able to *prove* additional properties and, for convenience in referring to the axioms and the properties that we are able to prove, we will adopt the following numbering scheme: The axioms will be numbered with capital Roman numerals, I, II, III, . . . , and the properties that we prove will be numbered with lower case Roman numerals, i, ii, iii, iv, ... It is important to *memorize* these axioms and to be able to associate each with its name. We now state the first four axioms.

AXIOM I. *The closure axiom.* If a and b are any natural numbers, then $a + b$ and $a \cdot b$ are unique natural numbers.

AXIOM II. *The commutative axiom.* If a and b are any natural numbers, then $a + b = b + a$ and $a \cdot b = b \cdot a$.

AXIOM III. *The associative axiom.* If a, b, c are any natural numbers, then $(a + b) + c = a + (b + c)$ and $(a \cdot b) \cdot c = a \cdot (b \cdot c)$.

AXIOM IV. *The distributive axiom for multiplication with respect to addition.* If a, b, c are any natural numbers, then

$$a \cdot (b + c) = (a \cdot b) + (a \cdot c).$$

In Axioms III and IV the parentheses play the role of *punctuation marks.* For example, $(a + b) + c$ indicates that we are to first find the sum of a and b and then find the sum of $(a + b)$ and c, while $a + (b + c)$ indicates that we are to first find the sum of b and c and then find the sum of a and $(b + c)$. As an illustration of the associative axiom for addition we have

$$(3 + 5) + 2 = 3 + (5 + 2),$$

since

$$(3 + 5) + 2 = 8 + 2 = 10 \quad \text{and} \quad 3 + (5 + 2) = 3 + 7 = 10.$$

When we write $a \cdot (b + c)$, as in Axiom IV, we indicate that b and c are to be added first, and then a is to be multiplied by the sum of b and c; for example

$$3 \cdot (5 + 2) = 3 \cdot 7 = 21,$$

where $3 \cdot (5 + 2)$ is read "three times the quantity five plus two." On the other hand, when we write

$$(a \cdot b) + (a \cdot c),$$

we indicate that the two multiplications are to be done first and that the products obtained are then to be added; for example

$$(3 \cdot 5) + (3 \cdot 2) = 15 + 6 = 21.$$

It is common practice to omit the parentheses in $(a \cdot b) + (a \cdot c)$ and write $a \cdot b + a \cdot c$, where, by agreement, we understand that the two multiplications are to be done first.

Addition is what we call a "binary" operation; that is, it is an operation we perform with *two* natural numbers. When adding a row of three or more natural numbers we perform a sequence of additions, each involving just *two* natural numbers. For example, in computing $2 + 5 + 8$ we add the 2 and the 5, to obtain 7, and then we add 7 and 8, to obtain 15. We progress to the sum 15 in *two* steps:

$$(2 + 5) + 8 = 7 + 8 = 15.$$

We could, of course, have obtained the sum 15 by writing

$$2 + (5 + 8) = 2 + 13 = 15.$$

Strictly speaking, if a, b, and c are natural numbers, the expression $a + b + c$ *has no meaning* unless we give it meaning (unless we define it), for the expression $a + b + c$ calls for finding the sum of *three* natural numbers—a "ternary" operation. With

$$a, b, c, d, e, \ldots$$

representing natural numbers, we *define*

$$a + b + c \quad \text{to mean} \quad (a + b) + c$$

and, having defined $a + b + c$, we *define*

$$a + b + c + d \quad \text{to mean} \quad (a + b + c) + d$$

and, having defined $a + b + c + d$, we *define*

$$a + b + c + d + e \quad \text{to mean} \quad (a + b + c + d) + e,$$

and so on.

It is important for us to realize that we compute $2 + 5 + 8$ by writing

$$(2 + 5) + 8 = 7 + 8 = 15,$$

because, *by definition*,

$$2 + 5 + 8 = (2 + 5) + 8.$$

Note also that we could just as well have defined $a + b + c$ to mean $a + (b + c)$ since, by the associative axiom for addition,

$$(a + b) + c = a + (b + c).$$

Suppose that a, b, c, and d are natural numbers and consider the expression $a + b + c + d$, which we have *defined* to mean $(a + b + c) + d$. Now $(a + b + c) + d$ has meaning only because $a + b + c$ has first been *defined*; namely $a + b + c = (a + b) + c$. Therefore,

$$a + b + c + d \quad \text{means} \quad [(a + b) + c] + d,$$

where, to avoid confusion that might have resulted had we written $((a + b) + c) + d$, we have used brackets in place of the outer parentheses.

When adding a row (or column) of three or more natural numbers, we are accustomed to taking such liberties as rearranging (commuting) the numbers to take advantage of combinations whose sum is ten. For example, confronted with the task of computing $6 + 7 + 4 + 3$, we might write $(6 + 4) + (7 + 3)$ or $(3 + 7) + (4 + 6)$. Of course, this is perfectly permissible, but we should be able to *prove* it. As a general example of this type let us *prove* that

$$a + b + c + d = (a + c) + (b + d).$$

Proof. Since

$$a + b + c + d = [(a + b) + c] + d \text{ (by definition)},$$

we are being asked to prove that the statement

$$[(a + b) + c] + d = (a + c) + (b + d)$$

is true; we proceed as follows:

(1) $[(a + b) + c] + d$ is a natural number,   (Why?)

(2) $[(a + b) + c] + d = [(a + b) + c] + d$,   (Why?)

(3) $(a + b) + c = a + (b + c)$,   (Why?)

$$(4) \quad b + c = c + b, \tag{Why?}$$

$$(5) \quad (a + b) + c = a + (c + b), \tag{Why?}$$

$$(6) \quad a + (c + b) = (a + c) + b, \tag{Why?}$$

$$(7) \quad (a + b) + c = (a + c) + b, \tag{Why?}$$

$$(8) \quad [(a + b) + c] + d = [(a + c) + b] + d, \tag{Why?}$$

$$(9) \quad [(a + c) + b] + d = (a + c) + [b + d]. \tag{Why?}$$

Therefore

$$(10) \quad [(a + b) + c] + d = (a + c) + (b + d). \tag{Why?}$$

Finally

$$(11) \quad a + b + c + d = (a + c) + (b + d). \tag{Why?}$$

Multiplication, also, is a "binary" operation; if $a, b, c, d, \ldots$ are natural numbers, the expression $a \cdot b \cdot c$ has no meaning unless we define it. We *define* $a \cdot b \cdot c$ to mean $(a \cdot b) \cdot c$ and, having defined $a \cdot b \cdot c$, we define $a \cdot b \cdot c \cdot d$ to mean $(a \cdot b \cdot c) \cdot d$, and so on. We now leave it to the reader to review the previous discussion of the associative law for addition of natural numbers and convince himself that a similar argument could be made with regard to the associative law for multiplication of natural numbers.

EXERCISE GROUP 2–1

Note: $a, b, c, d, e, \ldots$ represent natural numbers.

1. Prove that $a + b + c + d = [(b + d) + a] + c$.
2. Prove that $a + b + c = c + b + a$.
3. Prove that $a \cdot b \cdot c = b \cdot c \cdot a$.
4. Axiom IV is sometimes referred to as the "left-hand" distributive axiom. Prove the "right-hand" distributive axiom

$$(a + b) \cdot c = a \cdot c + b \cdot c.$$

5. Prove that

$$a \cdot c + c \cdot b = c \cdot (b + a).$$

6. Prove that

$$(a + b) \cdot (c + d) = a \cdot c + b \cdot c + a \cdot d + b \cdot d.$$

7. Prove that

$$a \cdot b + a \cdot c + a \cdot d = a \cdot (b + c + d).$$

The next three axioms to be examined are now given.

AXIOM V. *The cancellation axiom for addition. If a, b, c are any natural numbers such that $a + b = a + c$, then $b = c$.*

AXIOM VI. *The cancellation axiom for multiplication.* If a, b, c are any natural numbers such that $a \neq 0$ and $a \cdot b = a \cdot c$, then $b = c$.

AXIOM VII. *The identity axiom.* There exist *distinct* natural numbers 0 and 1 such that $a + 0 = a$ and $a - 1 = a$ for every natural number a.

Since addition of 0 or multiplication by 1 leaves any natural number unchanged, 0 and 1 are called "identity" elements of the natural number system; in particular, 0 is called the "additive identity" and 1 is called the "multiplicative identity."

Up to this point we have listed seven axioms which we *accept as being true*, and by using several of these axioms we are able to *prove* some additional properties of the natural number system.

(i) If a is any natural number, then $a \cdot 0 = 0$.

Property (i) may be proved as follows:

(1)	$a \cdot 0$ is a natural number,	(Ax. I)
(2)	$a \cdot 0 + 0 = a \cdot 0$,	(Ax. VII)
(3)	$a \cdot 0 = a \cdot 0 + 0$,	(Symmetry of =)
(4)	$0 + 0 = 0$,	(Ax. VII)
(5)	$a \cdot (0 + 0) = a \cdot 0 + 0$,	(Sub. principle)
(6)	$a \cdot (0 + 0) = a \cdot 0 + a \cdot 0$,	(Ax. IV)
(7)	$a \cdot 0 + a \cdot 0 = a \cdot 0 + 0$.	(Sub. principle)

Therefore

(8)	$a \cdot 0 = 0$.	(Ax. V)

(ii) If a and b are natural numbers such that $a \neq 0$ and $a \cdot b = 0$, then $b = 0$.

To prove (ii) we use (i), Axiom VI, and the substitution principle:

(1)	$a \cdot b = 0$,	(Given)
(2)	$a \cdot 0 = 0$,	[Prop. (i)]
(3)	$a \cdot b = a \cdot 0$.	(Sub. principle)

Therefore

(4)	$b = 0$.	(Ax. VI)

(iii) If a, b, c are natural numbers such that $a = b$, then

$$a + c = b + c \quad \text{and} \quad a \cdot c = b \cdot c.$$

The proof of (iii) requires only a straightforward application of the substitution principle:

(1)	$a + c = a + c$,	(Why?)
(2)	$a = b$.	(Given)

Therefore

$$(3) \quad a + c = b + c. \qquad \text{(Sub. principle)}$$

For the second part,

$$(1) \quad a \cdot c = a \cdot c, \qquad \text{(Why?)}$$

$$(2) \quad a = b. \qquad \text{(Given)}$$

Therefore

$$(3) \quad a \cdot c = b \cdot c. \qquad \text{(Sub. principle)}$$

EXERCISE GROUP 2–2

Note: $a, b, c, d, \ldots$ represent natural numbers.

1. Axiom V is sometimes referred to as the "left-hand" cancellation axiom for addition. *Prove* that if $a + c = b + c$, then $a = b$; that is, prove the "right-hand" cancellation axiom for addition.

2. Axiom VI is called the "left-hand" cancellation axiom for multiplication. *Prove* that if $a \cdot c = b \cdot c$ and $c \neq 0$, then $a = b$; that is, prove the "right-hand" cancellation axiom for multiplication.

3. Prove that if a is any natural number, then $0 + a = a$ and $1 \cdot a = a$; in other words, prove that 0 is a "left additive identity" and 1 is a "left multiplicative identity."

4. Prove that if $a + b = c + d$ and $a = c$, then $b = d$.

5. Prove that if $a = b$ and $c = d$, then $a + c = b + d$ and $a \cdot c = b \cdot d$.

6. Prove that $a + a = 2 \cdot a$. [*Hint:* Remember that $a = 1 \cdot a$ and use the distributive axiom.]

7. Prove that $a + a + a = 3 \cdot a$.

8. Using *only* the associative axiom for addition, the substitution principle, and the facts that

$$2 = 1 + 1, \quad 3 = 2 + 1, \quad 4 = 3 + 1,$$

prove carefully that $2 + 2 = 4$.

9. For any natural number a we *define* a^2 (a "squared") $= a \cdot a$. Prove that if $a = b$, then $a^2 = b^2$.

10. For any natural number a we *define* a^3 (a "cubed") $= a^2 \cdot a$. Prove that if $a = b$, then $a^3 = b^3$.

In our daily life we are constantly making such comparisons as "Joe is younger than Jim," "automobile X is less expensive to operate than automobile Y," or "Smith's yearly income is less than Johnson's." Each of these comparisons involves the concept of one natural number being "less than" another. In mathematics we use the symbol $<$ to mean "is less than" and we now state a precise definition of what we mean when we say that one natural number "is less than" another.

DEFINITION. If a and b are natural numbers, then we say that $a < b$ if, and only if, there is a natural number $c \neq 0$ such that $a + c = b$.

This common type of "if and only if" definition simply tells us that the statement $a < b$ is *true* if there is a natural number $c \neq 0$ such that $a + c = b$, and otherwise is *false*. Thus if a and b are natural numbers and we are given that $a < b$, we *know* that there is a natural number $c \neq 0$ such that $a + c = b$, for if this were not so, the statement $a < b$ would be false. On the other hand, if a, b, and $c \neq 0$ are natural numbers such that $a + c = b$, we know that $a < b$. We also write $b >$ (is "greater than") a if, and only if, $a < b$. We now see that one natural number may be related to another by being "less than," "greater than," or "equal to," and we state an axiom dealing with these relations.

AXIOM VIII. *The trichotomy axiom.* If a and b are any natural numbers, then *one* and *only one* of the statements $a < b$, $a = b$, $a > b$ is true.

By means of the trichotomy axiom we are now in a position to prove several additional properties of the natural number system.

(iv) If a and b are natural numbers such that $a \neq 0$ and $b \neq 0$, then $a \cdot b \neq 0$.

To prove (iv) we note that of the two statements $a \cdot b = 0$ and $a \cdot b \neq 0$, *one* and *only one* is true (why?) and we are to determine which. Let us assume, tentatively, that $a \cdot b = 0$ is the true statement. Then, on the basis of this assumption and the fact that $a \neq 0$, we must conclude by (ii) that $b = 0$. Since the conclusion that $b = 0$ has been reached by a valid application of (ii) but is nevertheless false, we must conclude that we have made the wrong choice; in short, $a \cdot b \neq 0$ is the true statement of the two, as we wished to show. The type of argument we have just employed in proving (iv) is called an *indirect proof*; to start with, we assumed false the statement we hoped eventually to prove true and, by showing that our assumption led us to a contradiction, we were able to reject that assumption in favor of the alternate one.

(v) If a is a natural number such that $a \neq 0$, then $0 < a$.

Property (v) is easily proved by simply noting that $0 + a = a$ (why?) so that $0 < a$ by definition of the relation $<$.

(vi) If a and b are natural numbers and $a \neq 0$, then $a + b \neq 0$.

We prove (vi) by considering two cases. First, we suppose $b = 0$, in which case $a + b = a + 0 = a$, so that $a + b \neq 0$. Next, we suppose that $b \neq 0$. In this case we note that one and only one of the statements $a + b = 0$ and $a + b \neq 0$ is true (why?), and, if we can identify the false statement, we will know which statement is true. If we tentatively assume $a + b = 0$ to be the true statement, the fact that $b \neq 0$ forces us to conclude that $a < 0$. This, of course, is a false conclusion, since we know by (v) that $0 < a$. Therefore, we must conclude that we have made the wrong choice; in other words, $a + b \neq 0$ is the correct statement, as we wished to show.

(vii) If a, b, c are natural numbers such that $a < b$, then $a + c < b + c$.

To prove (vii) we first think of the definition of the relation $<$; to show that $a + c < b + c$ we must show that there is some natural number, other than 0, which added to $a + c$ will yield $b + c$. We proceed as follows:

(1)	Since $a < b$, there is a natural number $d \neq 0$ such that $a + d = b$,	(Why?)
(2)	$(a + d) + c = b + c$,	[Prop. (iii)]
(3)	$(a + d) + c = a + (d + c)$,	(Ax. III)
(4)	$d + c = c + d$,	(Ax. II)
(5)	$(a + d) + c = a + (c + d)$,	(Sub. principle)
(6)	$a + (c + d) = (a + c) + d$,	(Ax. III)
(7)	$(a + d) + c = (a + c) + d$,	(Sub. principle)
(8)	$(a + c) + d = b + c$.	(Sub. principle)

Therefore

(9)	$a + c < b + c$.	(Why?)

In the above proof the reader should make certain that he understands the applications of the substitution principle.

(viii) If a, b, c are natural numbers such that $a < b$ and $c \neq 0$, then $a \cdot c < b \cdot c$.

To prove (viii) we must again think of the definition of the relation $<$; to prove that $a \cdot c < b \cdot c$ we must show that there is some natural number, other than 0, which added to $a \cdot c$ will yield $b \cdot c$. We proceed as follows:

(1)	Since $a < b$, there is a natural number $d \neq 0$ such that $a + d = b$,	(Why?)
(2)	$(a + d) \cdot c = b \cdot c$,	[Prop. (iii)]
(3)	$(a + d) \cdot c = a \cdot c + d \cdot c$,	(Why?)
(4)	$a \cdot c + d \cdot c = b \cdot c$,	(Sub. principle)
(5)	$d \cdot c \neq 0$.	[Prop. (iv)]

Therefore

(6)	$a \cdot c < b \cdot c$.	(Why?)

(ix) If a, b, c are natural numbers such that $a < b$ and $b < c$, then $a < c$.

The proof of (ix) follows from (vi); we argue as follows:

(1) Since $a < b$, there is a natural number
     $d \neq 0$ such that $a + d = b$,   (Why?)

(2) since $b < c$, there is a natural number
     $e \neq 0$ such that $b + e = c$,   (Why?)

(3) $(a + d) + e = c$,   (Sub. principle)

(4) $(a + d) + e = a + (d + e)$,   (Ax. III)

(5) $a + (d + e) = c$,   (Sub. principle)

(6) $d + e \neq 0$.   [Prop. (vi)]

Therefore

(7) $a < c$.   (Why?)

In proving (ix) we proved that the relation $<$ is "transitive." Property (ix) simply says: If a first natural number is less than a second and the second natural number is less than a third, then the first natural number is less than the third. Note that if, in this statement, we replace the phrase " is less than," *everywhere* it occurs, by the phrase "is equal to," we obtain the statement of the already familiar "transitive" property of the relation $=$. We have said that the relation $=$ is reflexive (that is, for any natural number a it is true that $a = a$) and symmetric (for any natural numbers a and b such that $a = b$ it is true that $b = a$). We leave these questions as exercises: For any natural number a is it true that $a < a$? For any natural numbers a and b such that $a < b$ is it true that $b < a$? In other words, is the relation $<$ reflexive and symmetric?

EXERCISE GROUP 2–3

Note: a, b, c, . . . represent natural numbers.

1. Prove that if $a < b$ and $c < d$, then $a + c < b + d$. [*Hint:* Use Properties (vii) and (ix).]

2. Prove that if $a < b$ and $c < d$, then $a \cdot c < b \cdot d$. [*Hint:* Consider the cases (1) $c = 0$ and (2) $c \neq 0$; use Properties (viii) and (ix).]

3. Prove that $1 < 2$.

4. Prove that if $1 < a$, then $a < a^2$. [*Hint:* First prove $a \neq 0$.]

5. Given $a < b$, prove that there is one *and only one* natural number $c \neq 0$ such that $a + c = b$.

We now consider one final axiom, *the axiom of induction.*

AXIOM IX. Let n be any natural number other than 0 and let $P(n)$, read "capital P of n," represent an assertion about the natural number n. If (1) the assertion $P(1)$ is true, and if (2) for any natural number n for which $P(n)$ is true, $P(n + 1)$ is also true, then *all* the assertions $P(1)$, $P(2)$, $P(3)$, ... are true. In other words, the assertion $P(n)$ is true for *any* natural number $n \neq 0$.

This axiom often doesn't make sense when one encounters it for the first time; therefore to help in understanding how the axiom is applied we consider a purely imaginary situation. Imagine an infinitely long ladder reaching into the sky, and imagine that its rungs are numbered 1, 2, 3, 4, ... starting with the bottom rung. Now suppose we are asked to *prove* that we can climb the ladder, where "climb the ladder" means "climb to *any* height." This, at first, appears to be an impossible task, but our difficulties will be surmounted easily by application of the axiom of induction. The proof proceeds as follows:

(1) Let $P(n)$ denote the statement "We can reach the nth rung."
(2) The statement $P(1)$ (we can reach the first rung) is true. (We will have to concede this.)
(3) If $P(n)$ is true, then $P(n + 1)$ is also true (if, for any natural number n, we can reach the nth rung, we can climb one rung higher and reach the $(n + 1)$th rung). (We will also have to concede that this is true.)
(4) Therefore, since we can reach the first rung, we can reach the second; since we can reach the second rung, we can reach the third; since we can reach the third rung, we can reach the fourth; and so on, indefinitely. In other words, we can climb to *any* height; therefore the statement $P(n)$ is true for *any* natural number $n \neq 0$.

We now find a more practical application of the axiom of induction in connection with the study of *exponents*. First let us consider the following definition.

DEFINITION. If a is any natural number, we define $a^1 = a$ (a^1 is read "a to the first power"), and if $n \neq 0$ is a natural number, we define $a^{n+1} = a^n \cdot a$.

This definition may appear not to define the meaning of a^n (a to the nth power) but it actually does. Note that a^1 is defined so that $a^2 = a^1 \cdot a$ is defined, a^2 is defined so that $a^3 = a^2 \cdot a$ is defined, a^3 is defined so that $a^4 = a^3 \cdot a$ is defined, and so on. A definition such as this is called a *recursive* definition and we say that we have defined a^n recursively. With this definition in mind we are now able, by the use of Axiom IX, to prove the following theorem.

THEOREM. If a, b, n are natural numbers such that $n \neq 0$, then $(a \cdot b)^n = a^n \cdot b^n$.

Proof.

(1) Let $P(n)$ denote the statement $(a \cdot b)^n = a^n \cdot b^n$.

(2) $P(1)$ is true. By the previous definition, $(a \cdot b)^1 = a \cdot b$, $a^1 = a$, $b^1 = b$, so that, by applying the substitution principle to the statement $a \cdot b = a \cdot b$, we obtain $(a \cdot b)^1 = a^1 \cdot b^1$.

(3) Assume that $P(n)$ is true for *some* n; that is, assume

$$(a \cdot b)^n = a^n \cdot b^n$$

is true for some particular natural number n. Then, by Property (iii), it is true that $(a \cdot b)^n \cdot (a \cdot b) = (a^n \cdot b^n) \cdot (a \cdot b)$. Now, by using Axioms II and III together with the previous definition, it is easily shown (do it!) that

$$(a^n \cdot b^n) \cdot (a \cdot b) = a^{n+1} \cdot b^{n+1},$$

and, also by the previous definition, $(a \cdot b)^n \cdot (a \cdot b) = (a \cdot b)^{n+1}$. Therefore, applying the substitution principle, we see that the statement $P(n+1)$, $(a \cdot b)^{n+1} = a^{n+1} \cdot b^{n+1}$, is true.

(4) We have shown two things: (a) $P(1)$ is true and (b) for any natural number n for which $P(n)$ is true, $P(n+1)$ is also true. Therefore, by Axiom IX, $P(n)$ is true for *any* natural number $n \neq 0$.

We may also prove the following theorem by using Axiom IX.

THEOREM. If a, m, n are natural numbers such that $m \neq 0$ and $n \neq 0$, then $a^m \cdot a^n = a^{m+n}$.

Proof.

(1) Let $m \neq 0$ be an arbitrary but fixed natural number and let $P(n)$ denote the statement $a^m \cdot a^n = a^{m+n}$.

(2) $P(1)$ denotes the statement $a^m \cdot a^1 = a^{m+1}$ and this statement is true by the definitions of a^{m+1} and a^1.

(3) Assume that the statement $P(n)$ is true for *some* n; that is, assume $a^m \cdot a^n = a^{m+n}$ is true for some particular natural number n. Then, by Property (iii), it is true that $(a^m \cdot a^n) \cdot a = a^{m+n} \cdot a$. Now, by using Axiom III together with the definition of a^{n+1}, we see that

$$(a^m \cdot a^n) \cdot a = a^m \cdot (a^n \cdot a) = a^m \cdot a^{n+1},$$

and furthermore, by the definition of $a^{(m+n)+1}$ together with Axiom III, we see that

$$a^{m+n} \cdot a = a^{(m+n)+1} = a^{m+(n+1)}.$$

Therefore, applying the substitution principle, we see that the statement $P(n + 1)$, $a^m \cdot a^{n+1} = a^{m+(n+1)}$, is true.

(4) We have shown two things: (a) $P(1)$ is true, and (b) for any natural number n for which $P(n)$ is true, $P(n + 1)$ is also true. Therefore, by Axiom IX, $P(n)$ is true for any natural number $n \neq 0$.

In expressions such as a^4, b^n, c^k (read "a to the 4th power," "b to the nth power," "c to the kth power"), 4, n, and k are called *exponents*, it being understood that a, b, c, n, k are natural numbers and that n and k are nonzero. An expression such as a^n means something to us so long as a and n are natural numbers and $n \neq 0$, but a^0 ("a to the zeroth power") has no meaning, for it has never been defined. We have complete freedom to define a^0 in any way we wish. However, we will want our definition to extend our earlier definition: $a^1 = a$ and $a^{n+1} = a^n \cdot a$ for any natural number $n \neq 0$. In other words, we will want to define a^0 in such a way that $a^{0+1} = a^0 \cdot a^1$. Let us assume for the moment that a^0 *is* defined and that $a^{0+1} = a^0 \cdot a$ *is* true. Then, since $a^{0+1} = a^1 = a$ and $a = 1 \cdot a$, it is also true that $1 \cdot a = a^0 \cdot a$. If $a \neq 0$, we may apply Axiom VI to obtain $1 = a^0$. Thus we are led to the following definition of the zero exponent.

DEFINITION. If a is a natural number such that $a \neq 0$, we define $a^0 = 1$.

We immediately note that if $a \neq 0$, the law $a^m \cdot a^n = a^{m+n}$ holds if either or both of the exponents m and n are zero. We shall leave the expression 0^0 undefined.

EXERCISE GROUP 2–4

Note: a, b, c, . . . represent natural numbers.

1. If $a < b$, prove by mathematical induction that $a^n < b^n$ for any natural number $n \neq 0$.

2. Prove Property (iii) by using mathematical induction.

3. The odd natural numbers are 1, 3, 5, 7, . . . , and if $n \neq 0$, the nth odd natural number is $2 \cdot n - 1$. Use mathematical induction to prove that the sum of the first n odd natural numbers is n^2; that is, prove that

$$1 + 3 + 5 + \cdots + (2 \cdot n - 1) = n^2.$$

4. We have defined $a^0 = 1$, provided $a \neq 0$. Can you think of any reason why we did not define 0^0?

5. If $n \neq 0$, prove that $0^n = 0$.

2–3 Using the axioms. Of the nine axioms that we have studied in the previous section, Axioms I through IV are most frequently used, but we often employ these axioms without realizing it. For example, if two students obtain different answers upon adding or multiplying two natural numbers, they know that at least one of the answers must be incorrect, but seldom are they aware that it is Axiom I that tells them this! The role played by Axioms II, III, and IV in the addition and multiplication of natural numbers is usually not appreciated, since addition and multiplication are learned by means of algorithms (rules for computation), which obscures the fact that these axioms are being used. For example, in computing $58 + 65$ by the addition algorithm, we write

$$
\begin{array}{r}
1 \\
58 \\
+65 \\
\hline
123
\end{array}
$$

where, on the surface, it does not appear that Axioms II, III, and IV have been used. However, if we understand that 58 means $5 \cdot 10 + 8$, 65 means $6 \cdot 10 + 5$, and 123 means $1 \cdot 100 + 2 \cdot 10 + 3$, and if we know the powers of ten ($10^2 = 100$, $10^3 = 1000, \ldots$, and so on) we can, just by knowing the addition and multiplication tables through $9 + 9$ and $9 \cdot 9$, add or multiply any two natural numbers without using the algorithms! We illustrate this fact by finding $58 + 65$; the reader should supply the justification for each step and identify the step in which we "carry."

$$
\begin{aligned}
58 + 65 &= (5 \cdot 10 + 8) + (6 \cdot 10 + 5) \\
&= 5 \cdot 10 + [8 + (6 \cdot 10 + 5)] \\
&= 5 \cdot 10 + [(6 \cdot 10 + 5) + 8] \\
&= 5 \cdot 10 + [6 \cdot 10 + (5 + 8)] \\
&= (5 \cdot 10 + 6 \cdot 10) + (8 + 5) \\
&= (5 + 6) \cdot 10 + 13 \\
&= 11 \cdot 10 + (10 + 3) \\
&= (11 \cdot 10 + 10) + 3 \\
&= (11 \cdot 10 + 1 \cdot 10) + 3 \\
&= (11 + 1) \cdot 10 + 3 \\
&= [(10 + 1) + 1] \cdot 10 + 3 \\
&= [10 + (1 + 1)] \cdot 10 + 3 \\
&= [10 + 2] \cdot 10 + 3 \\
&= [100 + 2 \cdot 10] + 3 \\
&= [1 \cdot 100 + 2 \cdot 10] + 3 \\
&= 1 \cdot 100 + 2 \cdot 10 + 3 \\
&= 123.
\end{aligned}
$$

In this lengthy computation we have taken some liberties that the reader may question, and we have not indicated how we arrived at any of the statements. Without omissions, the above development would start as follows:

(1) $58 + 65 = 58 + 65$. Since $58 = 5 \cdot 10 + 8$ and $65 = 6 \cdot 10 + 5$, we may substitute in (1) to obtain

(2) $58 + 65 = (5 \cdot 10 + 8) + (6 \cdot 10 + 5)$. Since

$$(5 \cdot 10 + 8) + (6 \cdot 10 + 5) = 5 \cdot 10 + [8 + (6 \cdot 10 + 5)],$$

we may substitute in (2) to obtain

(3) $58 + 65 = 5 \cdot 10 + [8 + (6 \cdot 10 + 5)]$, and so on.

As we become accustomed to working with the substitution principle we may feel free to take such short cuts, but we must, of course, be aware of all omissions. Returning to the computation itself, we note that the addition algorithm makes it possible to avoid the many steps taken above. Note, however, that the use of the axioms has *not really* been avoided and that the addition algorithm "works" *because* of the axioms and *because* of the fact that each digit of a natural number written in Hindu-Arabic notation has a "place value."

To illustrate how Axioms II, III, and IV apply in multiplying natural numbers, we consider $23 \cdot 34$. Using the multiplication algorithm, we write

$$
\begin{array}{r}
1 \\
23 \\
\times 34 \\
\hline
1\ 92 \\
69 \\
\hline
782
\end{array}
$$

where, again, it is not apparent that our axioms have been used. Now consider this same computation as it is carried out without using the algorithms (justify each step!):

$$
\begin{aligned}
23 \cdot 34 &= (20 + 3) \cdot (30 + 4) \\
&= (20 + 3) \cdot 30 + (20 + 3) \cdot 4 \\
&= (20 \cdot 30 + 3 \cdot 30) + (20 \cdot 4 + 3 \cdot 4) \\
&= [(2 \cdot 10) \cdot (3 \cdot 10) + 3 \cdot (3 \cdot 10)] + [(2 \cdot 10) \cdot 4 + 3 \cdot 4].
\end{aligned}
$$

At this point, we make the following observations:

$$(2 \cdot 10) \cdot (3 \cdot 10) = 6 \cdot 100, \quad 3 \cdot (3 \cdot 10) = 9 \cdot 10,$$

and
$$(2 \cdot 10) \cdot 4 = 8 \cdot 10. \quad \text{(Why?)}$$

Thus, upon substituting, we obtain

$$
\begin{aligned}
23 \cdot 34 &= [6 \cdot 100 + 9 \cdot 10] + [8 \cdot 10 + 3 \cdot 4] \\
&= [6 \cdot 100 + 9 \cdot 10] + [8 \cdot 10 + 12] \\
&= [6 \cdot 100 + 9 \cdot 10] + [8 \cdot 10 + (10 + 2)] \\
&= [6 \cdot 100 + 9 \cdot 10] + [(8 \cdot 10 + 1 \cdot 10) + 2] \\
&= [6 \cdot 100 + 9 \cdot 10] + [(8 + 1) \cdot 10 + 2] \\
&= [6 \cdot 100 + 9 \cdot 10] + [9 \cdot 10 + 2] \\
&= ([6 \cdot 100 + 9 \cdot 10] + 9 \cdot 10) + 2 \\
&= (6 \cdot 100 + [9 \cdot 10 + 9 \cdot 10]) + 2 \\
&= (6 \cdot 100 + [9 + 9] \cdot 10) + 2 \\
&= (6 \cdot 100 + 18 \cdot 10) + 2 \\
&= (6 \cdot 100 + [10 + 8] \cdot 10) + 2 \\
&= (6 \cdot 100 + [10 \cdot 10 + 8 \cdot 10]) + 2 \\
&= (6 \cdot 100 + [100 + 8 \cdot 10]) + 2 \\
&= ([6 \cdot 100 + 1 \cdot 100] + 8 \cdot 10) + 2 \\
&= ([6 + 1] \cdot 100 + 8 \cdot 10) + 2 \\
&= (7 \cdot 100 + 8 \cdot 10) + 2 \\
&= 7 \cdot 100 + 8 \cdot 10 + 2 \\
&= 782.
\end{aligned}
$$

We are indeed thankful for the convenience of the multiplication algorithm but, as with the addition algorithm, it is our axioms that make it work.

2–4 Subtraction and division. At this point, recalling that "subtractions" and "divisions" were performed in elementary arithmetic, you may wonder why discussion of these concepts has been avoided. We have not mentioned subtraction and division for the simple reason that they are not, in the strict sense, "operations" on the natural numbers and, before defining them, we shall need the following theorem.

THEOREM. Let a and b be natural numbers. Then, (1) there is *at most one* natural number c such that $a = b + c$, and (2), if $b \neq 0$, there is *at most one* natural number c such that $a = b \cdot c$.

The proof of this theorem is left as an exercise. Note that the theorem does *not* say that a natural number c exists in either case; the theorem *does* say that *if* a natural number c exists, then it is unique. We now define subtraction.

DEFINITION. If a and b are natural numbers, we define $a - b$ (read a "minus" b) to be the unique natural number c, if it exists, such that $a = b + c$.

The key phrase in this definition is "if it exists." Since there is a unique natural number, namely 3, such that $8 = 5 + 3$, $8 - 5$ is a natural number and we may write $8 - 5 = 3$. On the other hand, $4 - 9$ is not a natural number, for there is no natural number which may be added to 9 to yield 4. In other words, $a - b$ is a natural number for certain choices of natural numbers a and b, but for other choices of a and b the expression $a - b$ is not a natural number; in short, the natural numbers are not "closed" under subtraction. Similar remarks apply to division, which we now define.

DEFINITION. If a and b are natural numbers such that $b \neq 0$, we define $a \div b$ (read a "divided by" b) to be the unique natural number c, if it exists, such that $a = b \cdot c$.

Since there is a unique natural number, namely 4, such that $12 = 3 \cdot 4$, $12 \div 3$ is a natural number and we may write $12 \div 3 = 4$. However, $8 \div 5$ is not a natural number, for there is no natural number which may be multiplied by 5 to yield 8; the natural numbers are not "closed" under division. Division of a natural number by 0 is not defined; to show why we consider two cases.

Case 1. Consider the expression $a \div 0$, where $a \neq 0$. If we choose *any* natural number b and write $a \div 0 = b$ we see that this can *never* be true since, for *any* choice of b, we will have $0 = 0 \cdot b$ instead of $a = 0 \cdot b$.

Case 2. Consider the expression $0 \div 0$. If we choose *any* natural number b and write $0 \div 0 = b$ we see that this will *always* be true for *any* choice of b, since $0 = 0 \cdot b$.

Thus, by considering the above two cases, we see that it is not possible to define $a \div 0$ when $a \neq 0$ without contradicting Property (i), and it is not possible to define $0 \div 0$ *uniquely* since, by Property (i), $0 = 0 \cdot b$ for *every* natural number b.

We may now ask under what condition it is possible to subtract one natural number from another. To answer this question, let us first suppose that a, b, c are natural numbers such that $a - b = c$. Then, by definition, c is the natural number such that $a = b + c$. Thus if $c \neq 0$, it follows that $b < a$, and if $c = 0$, $b = a$. Therefore, if $a - b$ is a natural number, it must be true that $b \leq a$ ($\leq$ is read "is less than or equal to"). Let us now suppose that a and b are natural numbers such that $b \leq a$.

If $b < a$ there is a natural number $c \neq 0$ such that $b + c = a$ and $a - b = c$, by definition. If $b = a$ it is true that $b + 0 = a$, so that, by definition, $a - b = 0$. Therefore, if $b \leq a$, it must be true that $a - b$ is a natural number. We have now proved the following theorem.

THEOREM. If a and b are natural numbers it is possible to subtract b from a if, and only if, $b \leq a$.

2–5 Arithmetic in other bases. We are familiar with the fact that each digit of a natural number has a "place value." For example, in the natural number 253, 3 is called the 1's digit (units digit), 5 is called the 10's digit, and 2 is called the 100's digit; 253 means $2 \cdot 100 + 5 \cdot 10 + 3 \cdot 1$. Similarly, 5890 means

$$5 \cdot 1000 + 8 \cdot 100 + 9 \cdot 10 + 0 \cdot 1$$

or

$$5 \cdot 10^3 + 8 \cdot 10^2 + 9 \cdot 10^1 + 0 \cdot 1.$$

Remembering that we have defined $a^0 = 1$ for any natural number $a \neq 0$, we may write

$$253 = 2 \cdot 10^2 + 5 \cdot 10^1 + 3 \cdot 10^0,$$

and

$$5890 = 5 \cdot 10^3 + 8 \cdot 10^2 + 9 \cdot 10^1 + 0 \cdot 10^0.$$

We infer from these examples that, in general, each digit of a natural number is a multiplier of some power of ten, and that any natural number may be written as a sum of multiples of powers of ten, where each multiplier is a natural number less than ten. Because of this, we call our natural number system a *decimal* or *base-ten* system. No one knows for certain, but we suspect that we are today using a base-ten natural number system because of the fact that human beings have ten fingers; in short, it is natural for humans to count by tens. We also suspect that if humans had, say, six fingers, we might today be using a base-six natural number system. All this, of course, is mere speculation, but we would like to know whether, if forced to do so, we could learn to use a natural number system based on a natural number other than ten.

Suppose that an all-powerful dictator should decree that our decimal system be abolished and be replaced by the base-six system. If this happened, each one of us would have to learn immediately to write natural numbers in base-six notation and would have to memorize base-six addition and multiplication tables. First, let us learn how to write base-ten natural number symbols in base-six notation. Consider the following example.

EXAMPLE. Change 328 to base-six notation.

1. To "change 328 to base-six notation" means to write 328 as a sum of multiples of powers of six, each multiplier being a natural number less than six.

To accomplish this, we first write a table of powers of six, extending the table until we reach the highest power of six *not exceeding* 328. Thus

$$6^0 = 1, \quad 6^1 = 6, \quad 6^2 = 36, \quad 6^3 = 216.$$

We need not extend our table further, since $6^4 = 1296$ and all higher powers of six *exceed* 328.

2. Taking the highest power of six not exceeding 328 (in our case, 216), we find the *largest* whole number of times it is contained in 328. We see that 216 is contained in 328 *once*, with a remainder of 112, and we write

$$328 = 1 \cdot 216 + 112.$$

3. Again using our table of powers of six, we observe that the highest power of six not exceeding the remainder, 112, is 36, and that 36 is contained in 112 three times $(3 \cdot 36 = 108)$ with a remainder of 4. Thus we write

$$112 = 3 \cdot 36 + 4.$$

4. Substituting $3 \cdot 36 + 4$ for 112 in the equation $328 = 1 \cdot 216 + 112$, we obtain

$$328 = 1 \cdot 216 + 3 \cdot 36 + 4.$$

5. Finally, since $1 \cdot 216 = 1 \cdot 6^3$, $3 \cdot 36 = 3 \cdot 6^2$, $0 \cdot 6 = 0 \cdot 6^1$, and $4 \cdot 1 = 4 \cdot 6^0$, we may write

$$328 = 1 \cdot 6^3 + 3 \cdot 6^2 + 0 \cdot 6^1 + 4 \cdot 6^0.$$

Since each digit of a natural number written in base-six notation has a "place value," i.e., each digit is a multiplier of some power of six, we finally write 1304_6, read "one, three, zero, four, base six." The 6 written below and to the right of 1304 is called a *subscript* and we attach it only as a *reminder* that 1304 is a base-six number symbol—not a base-ten number symbol. As it turns out, then, $328 = 1304_6$; i.e., 328 and 1304_6 are merely different symbols for *the same* natural number. We have not attached a subscript to 328, and any natural number symbol hereafter written *without* a subscript should be *understood* to be a base-ten number symbol.

We should now be aware that any digit of a natural number symbol written in base-six notation is one of the symbols 0, 1, 2, 3, 4, 5—never any other—for the same reason that any digit of a natural number written in base-ten notation is one of the symbols 0, 1, 2, 3, 4, 5, 6, 7, 8, 9. If we should abandon the use of base-ten natural number symbols and adopt base-six symbols, we would abandon the symbols 6, 7, 8, and 9. We now write the base-ten and base-six natural number symbols side by side for the purpose of comparison:

Base ten: 0, 1, 2, 3, 4, 5, 6, 7, 8, 9, 10, 11, 12, 13, ...

Base six: 0, 1, 2, 3, 4, 5, 10, 11, 12, 13, 14, 15, 20, 21, ...

The subscript 6 has not been attached to 10, 11, 12, 13, ... in the base-six list because it is not needed; if we were dealing with these numbers out of context, we would use the subscript and write 10_6, 11_6, 12_6, and so on.

In order to be able to add and multiply with base-six natural number symbols, we need addition and multiplication tables:

+	0	1	2	3	4	5
0	0	1	2	3	4	5
1	1	2	3	4	5	10
2	2	3	4	5	10	11
3	3	4	5	10	11	12
4	4	5	10	11	12	13
5	5	10	11	12	13	14

×	0	1	2	3	4	5
0	0	0	0	0	0	0
1	0	1	2	3	4	5
2	0	2	4	10	12	14
3	0	3	10	13	20	23
4	0	4	12	20	24	32
5	0	5	14	23	32	41

Since we do not have these tables memorized, we consult them freely as we carry out the following computations.

EXAMPLE.
$$
\begin{array}{r}
11 \\
1352_6 \\
+3423_6 \\
\hline
5215_6
\end{array}
$$

Note that we have used the addition algorithm in the usual way—only our addition table is different! To check our computation, we now do the same addition in base-ten notation. Changing 1352_6 and 3423_6 to decimal notation, we have

$$1352_6 = 1 \cdot 6^3 + 3 \cdot 6^2 + 5 \cdot 6^1 + 2 \cdot 6^0 = 216 + 108 + 30 + 2 = 356,$$

and

$$3423_6 = 3 \cdot 6^3 + 4 \cdot 6^2 + 2 \cdot 6^1 + 3 \cdot 6^0 = 648 + 144 + 12 + 3 = 807.$$

We add these base-ten number symbols in the usual way, and obtain

$$
\begin{array}{r}
1 \\
356 \\
+807 \\
\hline
1163
\end{array}
$$

If it is correct, when converted to base-ten notation, our sum in base-six notation should be 1163. Carrying out this conversion gives us

$$5215_6 = 5 \cdot 6^3 + 2 \cdot 6^2 + 1 \cdot 6^1 + 5 \cdot 6^0 = 1080 + 72 + 6 + 5 = 1163.$$

Our computation in base-six notation is apparently correct.

EXAMPLE.

$$
\begin{array}{r}
32 \\
32 \\
243_6 \\
\times 145_6 \\
\hline
2143_6 \\
1 \\
1500_6 \\
243_6 \\
\hline
45443_6
\end{array}
$$

To check our multiplication, we shall carry it out using base-ten notation. Changing to base-ten notation, we have

$$243_6 = 2 \cdot 6^2 + 4 \cdot 6^1 + 3 \cdot 6^0 = 72 + 24 + 3 = 99,$$

and

$$145_6 = 1 \cdot 6^2 + 4 \cdot 6^1 + 5 \cdot 6^0 = 36 + 24 + 5 = 65.$$

Multiplication of the base-ten number symbols 99 and 65 gives us

$$
\begin{array}{r}
4 \\
65 \\
\times 99 \\
\hline
585 \\
585 \\
\hline
6435
\end{array}
$$

Assuming that 6435 is the correct product, we convert 45443_6 to base-ten notation:

$$45443_6 = 4 \cdot 6^4 + 5 \cdot 6^3 + 4 \cdot 6^2 + 4 \cdot 6^1 + 3 \cdot 6^0$$
$$= 5184 + 1080 + 144 + 24 + 3 = 6435.$$

Our multiplication of base-six number symbols seems to be correct.

In an earlier example we converted 328 to base-six notation, obtaining 1304_6. We now study an easier and more direct way to accomplish this conversion. First we shall show *how* it is done and later explain *why* it yields the correct result.

EXAMPLE. Convert 328 to base-six notation.

1. We divide 328 by 6, recording both the quotient and remainder:

$$
\begin{array}{r}
2 \\
6\ \overline{)3\ 2\ 8} \\
5\ 4
\end{array}
\qquad \text{R} \quad 4.
$$

2. We divide the quotient 54 by 6, again recording the quotient and remainder:

$$
\begin{array}{r}
6\ \overline{)5\ 4} \\
9
\end{array}
\qquad \text{R} \quad 0.
$$

3. We divide the quotient 9 by 6, recording the quotient and remainder:

$$
\begin{array}{r}
6\ \overline{)9} \\
1
\end{array}
\qquad \text{R} \quad 3.
$$

4. We divide the quotient 1 by 6, recording the quotient and remainder:

$$
\begin{array}{r}
6\ \overline{)1} \\
0
\end{array}
\qquad \text{R} \quad 1.
$$

The process ends at this point since, if it were continued, all subsequent quotients and remainders would be 0. We now write our remainders in reverse order, and obtain 1304, the base-six equivalent of 328! Our work may be arranged in a more compact fashion as follows:

$$
\begin{array}{rl}
0 & \quad \text{R} \quad 1 \\
6\ \overline{)1} & \quad \text{R} \quad 3 \\
6\ \overline{)\ 9} & \quad \text{R} \quad 0 \\
6\ \overline{)\ 5\ 4} & \quad \text{R} \quad 4 \\
6\ \overline{)3\ 2\ 8} &
\end{array}
$$

$$328 = 1304_6.$$

Our question now is: *Why* does this process give the correct result? To explain, we observe once again that 328 and 1304_6 are different symbols for *the same* natural number or, simply, $328 = 1304_6$. We have said little about division except to define it and point out that division is not an operation on the natural numbers, so that, in what we do next, we shall be using only our previous experience as our guide; we shall accept it as being reliable. Our experience tells us that if two equal natural numbers are divided by the same natural number the quotients will be equal and the remainders will be equal; we shall apply this principle to the numbers 328 and 1304_6. We also believe this: If a base-ten natural number symbol is divided by the base (that is, by 10), the remainder will be the 1's digit; if the quotient is divided by 10, the remainder will be the 10's digit; and so on. For example,

$$
\begin{array}{rll}
0 & \text{R} \quad 5 & \text{(1000's digit)}\\[-2pt]
10 \,\overline{)\,5} & \text{R} \quad 9 & \text{(100's digit)}\\[-2pt]
10 \,\overline{)\,5\ 9} & \text{R} \quad 3 & \text{(10's digit)}\\[-2pt]
10 \,\overline{)\,5\ 9\ 3} & \text{R} \quad 2 & \text{(1's digit)}\\[-2pt]
10 \,\overline{)\,5\ 9\ 3\ 2}
\end{array}
$$

This same principle applies to natural number symbols having a base other than 10; in particular, it applies to 1304_6. If we divide 1304_6 by the base (that is, by 10_6), the remainder will be the 1's digit; if we divide the quotient by 10_6, the remainder will be the 6's digit; and so on.

$$
\begin{array}{rll}
0 & \text{R} \quad 1 & \text{(216's digit)}\\[-2pt]
10_6 \,\overline{)\,1} & \text{R} \quad 3 & \text{(36's digit)}\\[-2pt]
10_6 \,\overline{)\,1\ 3_6} & \text{R} \quad 0 & \text{(6's digit)}\\[-2pt]
10_6 \,\overline{)\,1\ 3\ 0_6} & \text{R} \quad 4 & \text{(1's digit)}\\[-2pt]
10_6 \,\overline{)\,1\ 3\ 0\ 4_6}
\end{array}
$$

Note that we *have* divided by six: $10_6 = 1 \cdot 6^1 + 0 \cdot 6^0$.

Thus, using the principle that division of equal natural numbers by the same natural number yields equal quotients and equal remainders, we expect successive division of 328 by 6 to yield as remainders the digits of 328's base-six equivalent. Let us apply this "short cut" to another conversion.

EXAMPLE. Convert 437 to base-five notation.

$$
\begin{array}{rll}
0 & \text{R} \quad 3 & \text{(125's digit)}\\[-2pt]
5 \,\overline{)\,3} & \text{R} \quad 2 & \text{(25's digit)}\\[-2pt]
5 \,\overline{)\,1\ 7} & \text{R} \quad 2 & \text{(5's digit)}\\[-2pt]
5 \,\overline{)\,8\ 7} & \text{R} \quad 2 & \text{(1's digit)}\\[-2pt]
5 \,\overline{)\,4\ 3\ 7}
\end{array}
$$

Therefore, $437 = 3222_5$.

To verify this result we convert 3222_5 back to base-ten notation:

$$
\begin{aligned}
3222_5 &= 3 \cdot 5^3 + 2 \cdot 5^2 + 2 \cdot 5^1 + 2 \cdot 5^0\\
&= 3 \cdot 125 + 2 \cdot 25 + 2 \cdot 5 + 2 \cdot 1\\
&= 375 + 50 + 10 + 2\\
&= 437,
\end{aligned}
$$

as we wished to show.

We conclude this section with a brief mention of the *binary* (base-two) notation for natural numbers, which uses only two symbols, 0 and 1. For purposes of comparison we write the decimal and binary natural number symbols side by side.

Base ten: 0, 1, 2, 3, 4, 5, 6, 7, 8, ...

Base two: 0, 1, 10, 11, 100, 101, 110, 111, 1000, ...

The binary system has the obvious advantage that only two symbols (0 and 1) are needed to write any natural number in binary notation. Furthermore, the addition and multiplication tables for this system are short and can be quickly memorized. They are

+	0	1
0	0	1
1	1	10

and

×	0	1
0	0	0
1	0	1

We illustrate the use of these tables with two examples.

EXAMPLE.

$$\begin{array}{r} 1\quad 1 \\ 110101_2 \\ +110001_2 \\ \hline 1001110_2 \end{array}$$

(Verify this computation by carrying it out in base-ten notation.)

EXAMPLE.

$$\begin{array}{r} 1101_2 \\ \times 1011_2 \\ \hline 1101_2 \\ 1101_2 \\ 1101_2 \\ \hline 10001111_2 \end{array}$$

(Verify this computation by carrying it out in base-ten notation.)

There is something about multiplication of binary numbers, as illustrated in our last example, that is worthy of special mention. Note that the only multiplication that is actually done is multiplication by 1 and this, essentially, requires only copying the number being multiplied. In the above example we merely need to copy the numbers 1101_2, 11010_2, 1101000_2 in a column and add. Thus, in carrying out a binary multiplication, we need copy our multiplicand only once (with the proper number of zeroes attached) for each 1 in the multiplier and then add the numbers we have copied. The advantages of the binary notation that we have mentioned make binary numbers especially useful for high-speed digital computing. First of all, the fact that at most *two* symbols are needed to express any natural number in binary notation makes it possible to use *two*-position (on-off) switches in the computer circuits. Second, the simplicity of the binary addition and multiplication tables makes storage of these tables in a computer's "memory" a relatively simple matter. An obvious disadvantage of the binary notation is the often unwieldy length of binary number symbols; for example, to express 78 in binary notation requires seven digits.

EXERCISE GROUP 2–5

1. Write each of the following natural numbers in base-ten notation:
   - (a) 5734_8
   - (b) 122102_3
   - (c) 5046_7
   - (d) 1001101_2
   - (e) 4343_5
   - (f) 888_9

2. Convert each of the following natural numbers as indicated:
   - (a) 325 to base five
   - (b) 6561 to base nine
   - (c) 76 to base three
   - (d) 63 to base two

3. (a) Construct base-five addition and multiplication tables which give sums through $4 + 4$ and products through $4 \cdot 4$.
   (b) Use the tables to compute $3012_5 + 421_5$ and $234_5 \cdot 42_5$.
   (c) Change the numbers in (b) to base-ten notation, carry out the indicated operations to obtain answers in base-ten notation, and convert your base-ten answers to base five for comparison with the answers already obtained in (b).

4. Carry out each of the following conversions:
   (a) 425_6 to base four; to base eight.
   (b) 3213_4 to base two; to base seven.

You may accomplish each of these conversions by using the short cut. For example, to convert 425_6 to base eight, divide successively by 12_6.

   (c) Perform each of the above conversions by converting *first* to base ten.

5. Find the smallest and the largest base-ten natural numbers whose binary representations have seven digits.

2–6 Structure and isomorphism. In Section 1–1 we remarked that we would be studying the *structure* of some familiar number systems. At this point we may say that we *have* studied the structure of the natural number system; at least we have studied the most important features of that structure. However, the question remains: What do we mean by "structure"? Let us say at the outset that the structure of the natural number system consists of those properties of the system which are *independent* of the meanings of the symbols 0, 1, 2, 3, ... Since it is difficult for us to think of the symbols 0, 1, 2, 3, . . . apart from their meanings, even though these meanings are not clearly understood, this last statement doesn't really explain what we mean when we speak of "the structure of the natural number system." As an aid to understanding, we shall study two "number systems" whose structures are easier to bring out in the open for examination.

The following facts about *even* and *odd* natural numbers are assumed to be familiar:

(1) The sum of any two *even* numbers is always an *even* number.

(2) The sum of any two *odd* numbers is always an *even* number.

(3) The sum of any *even* number and any *odd* number is always an *odd* number.

(4) The product of any two *even* numbers is always an *even* number.

(5) The product of any two *odd* numbers is always an *odd* number.

(6) The product of any *even* number and any *odd* number is always an *even* number.

By using known properties of the natural number system we could *prove* these facts, but we shall not take time to do so; the reader is urged to try proving them for himself. Letting O represent "an odd number," E "an even number," and with $+$ and $\cdot$ meaning ordinary addition and multiplication of natural numbers, we write the above six statements in abbreviated symbolic form:

(1) $E + E = E$,   (2) $O + O = E$,   (3) $E + O = O$,

(4) $E \cdot E = E$,   (5) $O \cdot O = O$,   (6) $E \cdot O = E$.

Remembering that the operations of addition and multiplication of natural numbers are commutative, we may add the symbolic statements

(7) $O + E = O$ and   (8) $O \cdot E = E$

to this list.

The information contained in these eight statements may be put in still more compact tabular form, as shown below.

$+$	E	O		$\times$	E	O
E	E	O	and	E	E	E
O	O	E		O	E	O

We have here an example of what is called a "finite" or "miniature" arithmetic. In fact, after we define "number system" and "number" in Chapter 4, we will realize that the symbols E and O, apart from their meanings, are indeed "numbers." In the discussion that follows, then, we will boldly think of E and O as "numbers," even though we do not at this time understand what right we have to do so. The above tables give us the information necessary to describe the *structure* of this miniature arithmetic. For example, these tables tell us immediately that both addition and multiplication are commutative ($E + O = O + E$ and $E \cdot O = O \cdot E$) and, although it is not immediately obvious that both addition and multiplication are associative and that multiplication is distributive with respect to addition, the tables do give us the information necessary to *prove* these facts.

The tables also tell us of other properties possessed by this miniature arithmetic, properties whose mention we postpone until Chapter 4. We now say this: The collection of all properties of our miniature arithmetic which can be discovered by using the tables alone is called the *structure* of the arithmetic; the properties which can be discovered by use of the tables alone are precisely those properties which do not depend on the meanings of the symbols E and O.

Again, it may be difficult to divorce the symbols E and O from the meanings we originally assigned to them and so, to help our understanding of the concept of *structure*, we introduce another miniature arithmetic.

Consider the boxes pictured in Fig. 2–1. At one end of each box there are two pushbuttons, a red (R) one and a green (G) one, while at the opposite end of each box there are two light bulbs, a red (R) one and a green (G) one. Inside of each box is a maze of wires, switches, and relays, together with a battery to energize the relays and light the bulbs. The circuit in each box is designed in such a way that a bulb may be lit only by the push of one button (R or G) followed by a second push of a button (R or G). We describe the behavior of these boxes by writing the following symbolic statements:

Addition box	Multiplication box
(1) $R + R = R$	(1) $R \cdot R = R$
(2) $R + G = G$	(2) $R \cdot G = R$
(3) $G + R = G$	(3) $G \cdot R = R$
(4) $G + G = R$	(4) $G \cdot G = G$

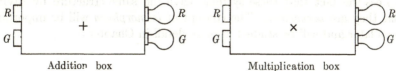

Addition box   Multiplication box

FIGURE 2–1

The statement $R + R = R$, for example, means "a push of the R button followed by a push of the R button lights the R bulb," and the statement $R + G = G$ means "a push of the R button followed by a push of the G button lights the G bulb."

In each of the symbolic statements on the left the symbol $+$ indicates that the statements apply to the "addition" box and, with this understanding, $+$ may be interpreted to mean "followed by," while the symbol $=$ may be interpreted to mean "lights." Similar interpretations may be given to each of the statements on the right except that the symbol $\cdot$ indicates that the statements apply to the "multiplication" box. We may condense the information contained in the above statements in tabular form:

$+$	R	G
R	R	G
G	G	R

and

$\times$	R	G
R	R	R
G	R	G

Forgetting about the meanings of the symbols R and G and concentrating only on the tables, we see that we have a "number system," i.e., another miniature arithmetic. Although the R-G tables have a vastly different origin than the E-O tables discussed earlier, we are impressed with the fact that they are very similar. In fact, if we take the E-O addition table and replace each E by R and each O by G, we obtain the R-G "addition" table; similarly, if we replace E by R and O by G in the E-O multiplication table, we obtain the R-G "multiplication" table. It is also true that replacing R by E and G by O in the R-G tables will yield the respective E-O tables. We say that the E-O and R-G systems have the *same structure*.

If the reader is disturbed by the fact that the R-G addition and multiplication tables seem artificial, he is missing the essential point of this discussion. First of all, a competent electrician *could* design and construct an "addition" and a "multiplication" box that would yield the above R-G addition and multiplication tables. Secondly, we have *purposely* considered these particular tables to show that it is possible for two number systems to have *the same structure* even though the symbols (numbers) of the two systems have different meanings. Mathematicians would express the fact that these systems have the same structure by saying that they are *isomorphic*. The concept of *isomorphism* will be important to us later and will be studied in more detail in Chapter 5.

CHAPTER 3

SETS, VARIABLES, AND STATEMENT FORMS

3–1 Sets. The concept of *set* is fundamental to the study of modern mathematics; in fact, mathematics has been called "the science of sets." However, we immediately encounter a disconcerting fact: we cannot explicitly define the noun "set." Everyone knows what a set is, but if we try to define it we find ourselves using such words as "class," "collection," "family," "aggregate," "assemblage," etc., all of which mean the same thing. In everyday conversation we use the idea of "set" whenever we speak of "a herd of cattle," "a string of fish," "a class of students," "a troop of boy scouts," or "a flock of sheep." We shall therefore have to start our study of modern mathematics by admitting into our thinking the concept of "set" with no further attempt to define it. As further examples of sets we might have:

> The set of University of Minnesota students.
> The set of members of your immediate family.
> The set of keys on your key ring.
> The set of books in your college library.
> The set of citizens of Minnesota.
> The set of representatives in the U.S. Congress.
> The set of all mathematics courses in the university catalogue.
> The set of words on this page.
> The set of families who reside in your block.
> The set of classes which meet at 8:30 a.m. on your campus.
> The set of all university student bodies in the U.S.

The objects that make up a set are called *elements* or *members* of the set. For example, each of you is an element of the set of students in your mathematics class. A given set is said to be "well defined" if, given any object, we can tell, at least in theory, whether or not it is a member of the given set. We shall not consider here such sets as "the set of all beautiful paintings," "the set of all loyal citizens," or "the set of all honest politicians," for these sets are certainly *not* well defined.

In writing about sets we shall represent each set either by a capital letter or by actually listing the members of the set, separated by commas, within braces. For example, we might denote "the set of all natural numbers less than 10" by A or some other capital letter of our choice, or we might actually write $\{0, 1, 2, 3, 4, 5, 6, 7, 8, 9\}$. In a discussion about this particular set, if we have specifically said "let A denote the set of all natural numbers less than 10," we can write

$$A = \{0, 1, 2, 3, 4, 5, 6, 7, 8, 9\}.$$

Similarly, we might denote "the set of all students in your mathematics class" by C, or we might actually list the total membership within braces as follows:

{Mary Jones, Jack Smith, Bill Johnson, . . .},

where the three dots . . . indicate that everyone concerned knows the membership of this set and could complete the list of members if asked to do so. Having specifically said "let C denote the set of students in your mathematics class" we may then write

$$C = \{\text{Mary Jones, Jack Smith, Bill Johnson, . . .}\}.$$

If we write the membership of any set within braces, the order in which we list the members is immaterial. For example, $\{3, 7, 9, 2, 4, 8, 0, 6, 1, 5\}$ is just as good a designation for "the set of all natural numbers less than 10" as is $\{0, 1, 2, 3, 4, 5, 6, 7, 8, 9\}$. If any set is designated by listing its members the important thing to remember is that each element must appear in the list *at least once*. It is customary to enter each element in the membership list *exactly once*. However, occasions will arise when it will be convenient for us to allow an element to be entered more than once in the membership list of a set. There is no harm in doing this as long as we remember, when counting the elements of the set, that we count each element but once, even though it may appear in the list more than once. Unless something is said to the contrary we shall assume that the membership list of any set we discuss contains each element of the set once and only once; that is, each membership list is a list of *distinct* elements.

We now introduce the notions of *finite* and *infinite* sets. We define a *finite* set as follows:

DEFINITION. A set is said to be *finite* if there is a natural number n such that the set has exactly n members.

A set that is *not* finite is said to be *infinite*. For example, the set of all natural numbers is infinite. In this text most of the sets we shall discuss will be finite. All of the sets we have mentioned so far are finite except the set of natural numbers, $\{0, 1, 2, 3, 4, 5, . . .\}$, which we will agree from now on to denote by N.

3–2 Subsets. The notion of a *subset* is a simple and obvious one:

DEFINITION. If A and B are any sets, A is said to be a *subset* of B if, and only if, each member of A is also a member of B.

Note the "any" and the "if and only if" in this definition. The definition says that if A is *any* set such that each member of A is also a member of B, then A is a subset of B and, also, if A is *any* set and B is *any* set such that A is a subset of B, then each member of A is a member of B. It immediately follows that *any set is a subset of itself.* Whenever we wish to indicate in writing that a set A is a subset of a set B we simply write $A \subset B$, which is read "A is a subset of B" or "A is contained in B." For example, if F is the set of all Fords on the campus parking lot and A is the set of all automobiles on the campus parking lot, then $F \subset A$; if M is the set of all men in your mathematics class and C is the set of all students in your mathematics class, then $M \subset C$; if B is the set of all books in the library and H is the set of all history books in the library, then $H \subset B$.

Another shorthand symbol that will be very useful is $\in$, which means "is a member of," "is an element of," or, simply, "belongs to." This symbol is used whenever we wish to indicate that some element "is a member of" or "belongs to" some particular set. For example, if C is the set of students in your mathematics class and Mary Jones is a student in your mathematics class, then we may write Mary Jones $\in C$.

There is an inexhaustible supply of sets about us, both real and imaginary, which may be used to illustrate properties of sets and relations between sets. However, we wish to draw *general* conclusions about sets, without being influenced by any knowledge we may have concerning their memberships. For this reason and, incidentally, because it is often inconvenient to list the members of sets close at hand, we shall use *abstract* sets in most of our discussions, i.e., sets whose members are represented by symbols. The symbols we shall most often use are the letters of the alphabet $(a, b, c, d, e, \ldots)$. In discussing such sets as $\{a, b, c, d\}$ and $\{p, q, r, x, u, z\}$, it is not important what objects the symbols $a, b, c, d, p, q, r, x, u, z$ represent.

Occasionally we will use subscript notation when we list the membership of a set. For example, a set having five elements might be written $\{a_1, a_2, a_3, a_4, a_5\}$ and a set having an arbitrary number, n, of elements might be written $\{a_1, a_2, a_3, \ldots, a_n\}$. The symbols $a_1, a_2, a_3, \ldots$ are read "a sub 1," "a sub 2," "a sub 3," and so on. The numbers $1, 2, 3, \ldots$ are here called *subscripts*, and they have no other function than to serve as labels—just as a railroad numbers its trains, an airline numbers its flights, or the warden of a prison numbers his prisoners.

In Section 3–1 we mentioned that we would be concerned only with "well-defined" sets. Sometimes it will be difficult to decide whether a certain set is "well defined" because we will not be able to agree on what elements should appear in its membership list. There is one set, however,

upon which we can all agree: the *empty* set, i.e., *the set with no members.*
This set will play such an important part in our discussions that we
reserve a special symbol for it, $\emptyset$.   The following are examples of the
empty set:

> The set of eggs in an empty basket.
> The set of coins in the pocket of a man who is broke.
> The set of billionaires in this classroom.
> The set of Maxwells on the campus parking lot.

Some may argue that these sets are different and that we have no right
to speak of "the" empty set but, rather, should speak of "an" empty set.
However, note that each of these sets has the same membership, *no members at all*, and that, although their descriptions are different, these
descriptions refer to the same set, namely $\emptyset$.   We shall *agree* that $\emptyset$ *is a
subset of every set.*   This convention seems to contradict our earlier definition of subset for, since $\emptyset$ has no members, how can each of its elements
belong to another set?   This is a good question!   Let it be said in answer
that it is *convenient* to be able to say that $\emptyset$ is a subset of every set and,
in addition, *it does no harm.*   Furthermore, we can argue that if A is *any*
set, $\emptyset$ has no element which is *not* a member of A.   Part of the convenience
of having $\emptyset$ as a subset of every set comes into play when we wish to count
the subsets of any given finite set.   We have the following theorem:

THEOREM.   If A is any set having n distinct members, then A has 2^n
distinct subsets.

Before trying to prove this theorem let us verify it for a few values of n.
The set $\{a\}$ having *one* member has the following subsets: $\emptyset$ (by agreement) and $\{a\}$, or two subsets in all. Since $2 = 2^1$, the theorem is verified
for $n = 1$. The set $\{a, b\}$ having *two* members has the following subsets:
$\emptyset$ (by agreement), $\{a\}$, $\{b\}$, and $\{a, b\}$, or four subsets. Since $4 = 2 \cdot 2 = 2^2$,
the theorem is verified for $n = 2$. The set $\{a, b, c\}$ having *three* members
has the following subsets: $\emptyset$, $\{a\}$, $\{b\}$, $\{c\}$, $\{a, b\}$, $\{a, c\}$, $\{b, c\}$, and
$\{a, b, c\}$, or eight subsets. Since $8 = 2 \cdot 2 \cdot 2 = 2^3$, the theorem is verified for $n = 3$. Finally, since every set is a subset of itself and since $\emptyset$
has no members, it is clear that $\emptyset$ has only *one* subset.   Thus, since
$1 = 2^0$, the theorem is verified for $n = 0$.

Our results so far, then, show us that the theorem is true for the cases
$n = 0, 1, 2$, and 3. This verification, however, does not *prove* the theorem.
For all we know the theorem may not be true if $n = 7$ or some other
value, and therefore we must prove it for the general case.   To do this,
let us consider the nonempty set

$$A = \{a_1, a_2, a_3, \ldots, a_n\}$$

1st toss   2nd toss   3rd toss

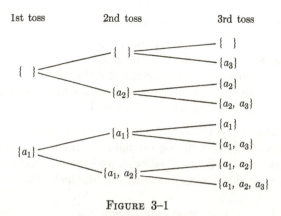

FIGURE 3–1

having n distinct members, and let us actually construct a subset of A. We first ask ourselves how we might choose the elements to be inserted within the braces which will contain the elements of our subset. One method of choosing the membership depends on pure chance—the coin-tossing method. We proceed as follows. We consider each of the elements $a_1, a_2, a_3, \ldots, a_n$ of A one at a time, and decide whether or not it is to be a member of our subset by tossing a coin. Let us agree that a "head" puts an element in the membership list and a "tail" leaves it out (of course, the reverse would be just as good).

We consider a_1 first and toss our coin so that after the first toss our subset will look like $\{\ \ \}$ or $\{a_1\}$. We consider a_2 next and toss our coin so that after the second toss our subset will look like $\{\ \ \}$, $\{a_1\}$, $\{a_2\}$, or $\{a_1, a_2\}$. We consider a_3 next and toss our coin so that after the third toss of the coin our subset will look like $\{\ \ \}$, $\{a_1\}$, $\{a_2\}$, $\{a_3\}$, $\{a_1, a_2\}$, $\{a_1, a_3\}$, $\{a_2, a_3\}$, or $\{a_1, a_2, a_3\}$. The possible results so far can be conveniently displayed in the form of a so-called "tree," as shown in Fig. 3–1. It now *appears* that it is possible to construct exactly 2^1 subsets of A by considering the element a_1, exactly 2^2 subsets of A by considering the elements a_1, a_2, exactly 2^3 subsets of A by considering the elements a_1, a_2, a_3, and so on.

Let us *assume* that it is possible to construct exactly 2^k subsets of A by choosing either none or some combination of one or more of the first k elements $a_1, a_2, a_3, \ldots, a_k$. If this is true, the $(k + 1)$th toss of the coin will either insert the element a_{k+1} in the membership list of each of these 2^k subsets or leave it out—in other words, the number of possible subsets will be *doubled*. Thus, by choosing either none or some combination of one or more of the elements

$$a_1, a_2, a_3, \ldots, a_k, a_{k+1},$$

exactly $2^k \cdot 2 = 2^{k+1}$ subsets of A can be constructed.

Let us now summarize what we have done.

(1) We have considered the statement $P(n)$: "A nonempty set A having n distinct members has 2^n distinct subsets."

(2) We have proved that the statement $P(1)$ is true.

(3) We have shown that if $P(n)$ is true for an arbitrary value of n ($n = k$), then $P(n)$ is true when $n = k + 1$.

Finally, then, we can conclude by the axiom of induction that the statement (theorem) is true for all nonzero values of n and, since we proved separately that $\emptyset$ has 2^0 subsets, we conclude that any set (empty or not) having n distinct members has 2^n distinct subsets.

EXERCISE GROUP 3–1

1. For each of the following categories, describe in words two sets whose individual members are (a) people, (b) books, (c) buildings, (d) letters of the alphabet. [*Example:* (a) "The set of all ballplayers on the Los Angeles Dodgers team."]

2. List within braces the membership of each of the following sets: (a) the set of the original thirteen colonies, (b) the set of distinct letters in your given name, (c) the set of U.S. presidents who have served three or more terms, (d) the set of natural numbers which are greater than 3 but less than 9.

3. We define $\not\subset$ to mean "is *not* a subset of" and we define $\notin$ to mean "is *not* an element of." Insert between each set or element in the left column and the corresponding set in the right column that one of the symbols $\subset$, $\in$, $\not\subset$, and $\notin$ which makes a correct statement:

a	$\{a, b, c\}$
e	$\{a, b, c\}$
2	$\{1, 2, 3, 4\}$
3	$\{1, 3\}$
$\{3\}$	$\{3\}$
3	$\{3\}$
$\{b, a\}$	$\{a, b\}$

4. If A, B, and C are sets such that $A \subset B$ and $B \subset C$, prove that $A \subset C$.

5. Given $A = \{a, b, c, d\}$. Write all sixteen subsets of A.

6. A hostess is planning a dinner party to which she is going to invite *at least two* guests, to be selected from a list of eight friends. In how many different ways can she do this? [*Hint:* Find the number of subsets having two or more members.]

7. A student is required to take exactly four courses of six that are offered. In how many different programs may the student enroll? [*Hint:* Find the number of four-element subsets.]

8. Santa Claus has five gifts in his bag. In how many ways can he give a certain child *at least one* gift?

Since we have used the symbol = only in passing, in connection with sets, we now give a formal definition:

DEFINITION. If A and B are sets, then the statement $A = B$ means precisely that A and B are symbols for *the same* set.

We will sometimes need to prove that two sets are equal, but unless their complete membership lists are available to us for examination, the task will be difficult. On the basis of the above definition we may prove the following theorem, which will be useful to us on occasion:

THEOREM. If A and B are any sets, then $A = B$ if, and only if, $A \subset B$ and $B \subset A$.

To prove this theorem we must prove (1) that if $A = B$, then $A \subset B$ and $B \subset A$, and (2) that if $A \subset B$ and $B \subset A$, then $A = B$. The proof of (1) is trivial: if $A = B$, it follows from the above definition that every element of A belongs to B and every element of B belongs to A. To prove (2) we simply note that the statement $A = B$ is either true or false. Then, supposing that $A \subset B$ and $B \subset A$, let us assume that the statement $A = B$ is *false*. If our assumption is correct, either A has at least one element not contained in B, B has at least one element not contained in A, or each set has at least one element not contained in the other. In any case, our hypothesis that $A \subset B$ and $B \subset A$ is contradicted and the statement $A = B$ must, therefore, be true.

Again, pay particular attention to the "any" and the "if and only if." On the basis of the above theorem we will know that two sets are "equal" or "the same" if we can show that each is a subset of the other, and we will know that two sets are *not* equal if we can show that one of them is *not* a subset of the other. Many situations will arise where we will be dealing with sets which are *notationally different* but are nevertheless "equal." As an example, consider the remark that was made earlier in regard to the ordering of the elements in the membership list of a set. The sets $\{0, 1, 2, 3, 4, 5, 6, 7, 8, 9\}$ and $\{3, 7, 9, 2, 4, 8, 0, 6, 1, 5\}$ are equal by our above theorem, since each element of either set also belongs to the other set; in short, each of these sets is a subset of the other. It has also been mentioned that we usually try to enter each element of a set once and only once in its membership list, but that no harm is done by entering an element more than once. As an illustration, consider the sets $\{p, q, r\}$ and $\{q, r, q, p, r, q\}$. In spite of the fact that q is entered three times and r is entered twice in the membership list of the second set, *by our theorem* these sets are equal: each element of $\{p, q, r\}$ belongs to, i.e., appears in the membership list of, $\{q, r, q, p, r, q\}$ and each element of $\{q, r, q, p, r, q\}$ belongs to $\{p, q, r\}$. We shall later encounter sets having elements which are entered in the membership list *infinitely many times*.

3–3 Variables and statement forms. In our study of sets, we frequently *make statements* about the members of a set, or perhaps prove, disprove, or merely interpret statements others have made about such members. In many of the statements we encounter we find one or more of the words "all," "any," "each," "every," or "some." These words are very helpful, for by using one or more of them we can often compress the sense of many, sometimes infinitely many, statements into a *single* statement. These *single* statements which convey the meaning of many statements will be the object of our study in this section.

Let us suppose that a certain college requires each of its freshmen to take a mathematics course and that

$$F = \{\text{Andrew Anderson, Betty Brown, Charles Cook, } \ldots ,$$
$$\text{William Wilson}\}$$

is the set of all freshmen enrolled at this particular college. Now let us consider the statement "*all* freshmen take mathematics," where it is understood that "all freshmen" means precisely those belonging to the set F. Note that this statement says the same thing as *all* of the statements

"Andrew Anderson takes mathematics,"

"Betty Brown takes mathematics,"

"Charles Cook takes mathematics,"

$\vdots$

"William Wilson takes mathematics."

The word "all" has certainly served a useful purpose here; we have been able to compress the sense of many statements into one statement. Of course, we would not have been able to do this unless we had a clear idea of the meaning of "all freshmen." Let us now consider this same statement as a mathematician might write it: "x takes mathematics, where $x \in F$," or "if $x \in F$, then x takes mathematics," or "x takes mathematics whenever $x \in F$."

In the statement "x takes mathematics, where $x \in F$," x is called a *variable* and "x takes mathematics" is called a *statement form*. Note that we have not defined "variable" or "statement form"; we have merely given examples. In fact, we shall leave these concepts undefined. However, as with the concept of "set," this will not prevent us from talking about variables and statement forms. In the statement form "x takes mathematics," the variable x plays exactly the same role as the blank in "_____ takes mathematics." "x takes mathematics" is *not* a statement and becomes a statement only when we specify to what set x belongs. The set F of all freshmen is called the *domain* of the variable

x and the statement form "x takes mathematics" accompanied by information giving the domain of x is called a statement; this is the same as
saying that "_____ takes mathematics" becomes a statement only
when we specify the set of names (the domain) from which *any* name may
be selected for insertion in the blank.

Many statements that we encounter involve two or more variables.
Suppose, for example, that the college mentioned earlier also requires
each freshman to take English and history. Then, if F is the set of
freshmen and S is the set of required subjects, the statement "all freshmen
take mathematics, English, and history" might be written: "x takes y,
where $x \in F$, $y \in S$." The word "any" appears nowhere in this statement
but it is *understood*; "where $x \in F$, $y \in S$" is understood to mean "where
x is *any* member of F and y is *any* member of S." The statement itself
means the same thing as the set of *all* statements that can be made by
selecting a member from F (a "value" of x) and a member from S (a
"value" of y). "x takes y" is a *statement form* involving two variables,
and "$x \in F$, $y \in S$" specifies the domains of these variables. In Chapter 2
we encountered many statements involving two or more variables, e.g.,
the commutative axiom: "If a and b are any natural numbers, then
$a + b = b + a$ and $a \cdot b = b \cdot a$." This statement might be written: "If
$a,b \in N$, then $a + b = b + a$ and $a \cdot b = b \cdot a$," where "$a + b = b + a$
and $a \cdot b = b \cdot a$" is the *statement form* and "$a,b \in N$" specifies the domains of the two variables. Since a and b have the same domain, we
have written "$a,b \in N$" rather than "$a \in N$, $b \in N$." As a further example, the statement "if $a,b,c \in N$, then $(a \cdot b) \cdot c = a \cdot (b \cdot c)$" is just
our familiar associative axiom for multiplication written in a more com-
pact way.

We need not revolutionize our way of making statements for purposes
of mathematical manipulation, but we do need to know how to interpret
mathematical statements which are written in abbreviated form. We
will encounter many compact statements throughout the remainder of
this text.

EXERCISE GROUP 3-2

1. Consider the following statement forms:

   (a) x had y for breakfast.
   (b) x bought x a ticket to the movies last night.
   (c) x weighs not over ten pounds more than y.
   (d) x and y received the same grade in course z.

For each of the statement forms above, specify sets that make sensible domains
for the variables involved.

2. In Chapter 2 the truth of eighteen statements about the natural numbers
was either assumed or proved. (a) Identify the *statement form* in each statement

and identify the part of statement that specifies the domains of the variables involved. (b) Write each statement in as compact a form as possible.

3. Compose three statement forms involving two variables, and specify sensible domains for the variables.

4. Compose three statement forms involving one variable occurring in more than one place, and specify a sensible domain for the variable.

3–4 Unions, intersections, differences, and products. In this section we shall discuss how we put sets together in order to construct new sets; several interesting and useful results will arise from this discussion.

One way of putting sets together to construct a new set is to form their *union*, and we now define this concept:

DEFINITION. The *union* of two or more sets is the set of all elements each of which is a member of *at least one* of the given sets.

The union of two sets, say A and B, is designated by $A \cup B$ which is read "A union B" or "the union of A and B." $A \cup B$, then, is the symbol for the set each of whose elements is a member of either A or B, or possibly both. For example, if $A = \{a, b, c\}$ and $B = \{a, c, e\}$, then $A \cup B = \{a, b, c, e\}$. It is important to note that when we list the membership of $A \cup B$ we insert those elements and only those elements which have the property of belonging to A, to B, or to both.

We may also put sets together by forming what we call their *intersection*, which we now define:

DEFINITION. The *intersection* of two or more sets is the set of all elements each of which is a member of every given set.

The intersection of two sets, say C and D, is designated by $C \cap D$ which is read "C intersection D" or "the intersection of C and D." $C \cap D$, then, is the symbol for the set each of whose elements is a member of *both* C and D. For example, if r, s, t, u, v are distinct elements, and if $C = \{r, s, t\}$, and $D = \{t, u, v\}$, then $C \cap D = \{t\}$, for t is the only element belonging to *both* C and D. Sets that have one or more members in common are said to be *overlapping* sets. Sets C and D in the example just considered are overlapping sets, since they have one member, t, in common. Sets that have no members in common are said to be *disjoint*. $E = \{a, b, c, g\}$ and $F = \{j, k, l\}$ are disjoint, and we note also that $E \cap F = \emptyset$, since E and F have no members in common. In fact, we might define disjoint sets as *sets whose intersection is* $\emptyset$.

The *difference* of two sets is defined as follows:

DEFINITION. If A and B are any sets, we define $A - B$, read "A minus B," to be the set of all elements each of which *is* a member of A but *is not* a member of B.

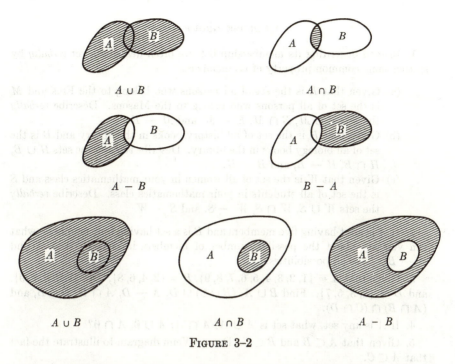

$A \cup B$ $A \cap B$

$A - B$ $B - A$

$A \cup B$ $A \cap B$ $A - B$

FIGURE 3-2

For example, if $A = \{a, b, c, d\}$ and $B = \{b, d\}$, then $A - B = \{a, c\}$; if $P = \{p, q, r\}$ and $Q = \{s, t, u, v\}$, then $P - Q = \{p, q, r\} = P$; and if $C = \{l, m\}$ and $D = \{j, k, l, m\}$, then $C - D = \emptyset$. The concept of "difference" will not play an important part in our later work but we will occasionally find it convenient to use.

To help us visualize sets and their unions, intersections, and differences, we often draw *Venn diagrams* in which sets are represented geometrically as "sets of points" enclosed by curves. Although the concept of "point," like the concept of "set," is undefined, in drawing Venn diagrams we rely on our geometric intuition. If we draw a closed curve it seems natural to talk about "the set of points" inside the curve, in spite of the fact that we are unable to define "set of points" explicitly. In the Venn diagrams of Fig. 3-2, A and B are sets of points inside closed curves; certain portions of each diagram have been shaded to illustrate $A \cup B$, $A \cap B$, $A - B$, $B - A$.

Since Venn diagrams deal with particular sets, they do not *prove* anything about sets in general; when we prove things about sets we want to know that our conclusions hold true for *any* sets. However, Venn diagrams are useful as an aid to our thinking. We use them as a man with a fractured leg uses a crutch; when the fracture is healed the crutch is thrown away.

1. Instead of writing its membership list, we often describe a set *verbally* by stating some common property of its members.

    (a) Given that E is the set of all persons who belong to the Elks and M is the set of all persons who belong to the Masons. Describe *verbally* the sets $E \cup M$, $E \cap M$, $E - M$, and $M - E$.

    (b) Given that H is the set of all history books in the library and B is the set of all biology books in the library. Describe *verbally* the sets $H \cup B$, $H \cap B$, $H - B$, and $B - H$.

    (c) Given that W is the set of all women in your mathematics class and S is the set of all students in your mathematics class. Describe *verbally* the sets $W \cup S$, $W \cap S$, $W - S$, and $S - W$.

2. If A is a set having five members and B is a set having four members, what can you say about the possible number of members in $A \cup B$, $A \cap B$, and $A - B$? Give all possibilities.

3. Given that $A = \{1, 2, 3, 4, 5, 6, 7, 8, 9\}$, $B = \{2, 4, 6, 8\}$, $C = \{1, 3, 5, 7, 9\}$, and $D = \{2, 3, 5, 7\}$. Find $B \cup A$, $(B \cup C) \cup D$, $A - D$, $A \cap D$, $D \cap B$, and $(A \cap B) \cap (C \cap D)$.

4. If A is any set, what set is $A \cup A$, $A \cap A$, $A \cup \emptyset$, $A \cap \emptyset$?

5. Given that $A \subset B$ and $B \subset C$. Draw a Venn diagram to illustrate the fact that $A \subset C$.

6. Draw a Venn diagram to illustrate the facts that $A \cap B \subset B$ and $A \subset A \cup B$.

7. Consider the Venn diagram of Fig. 3–3. Determine the portion of this diagram that should be shaded to illustrate each of the following sets: $A \cup C$, $A \cap B$, $(A \cup C) \cap B$, $(A \cup B) \cup C$, $(A \cap B) \cap C$, $(A \cap B) - C$, $(C \cap B) \cup (A \cap C)$.

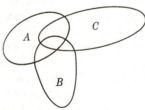

FIGURE 3–3

Having defined *union* and *intersection*, we shall now mention some of the immediate consequences of these definitions. First of all, if we form either the union or the intersection of *any* two well-defined sets, we *always* obtain some well-defined set. To see the significance of this remark, let us consider a set X and the "set of sets" S whose elements are all the subsets of X. Since it is often confusing to speak of a "set of sets," we

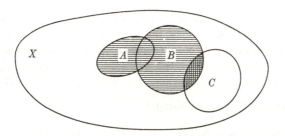

FIGURE 3–4

will agree to call such a set a "family" of sets. Several sets mentioned earlier are families of sets, i.e., have elements which are themselves sets. For example, "the set of words (sets of letters) on this page" and "the set of all university student bodies (sets of students) in the U.S." are such families.

If we take *any* two elements (sets) belonging to S and form their union, we always obtain some member of S. Similarly, if we take any two elements (sets) belonging to S and form their intersection we always obtain some member of S. The Venn diagram of Fig. 3–4 illustrates these facts. In the diagram, A, B, and C are pictured as subsets of X (members of S). The set $A \cup B$ indicated by the horizontal shading and the set $B \cap C$ indicated by the vertical shading are subsets of X, and hence members of S.

Although the concept of "an operation on a set" has not been explained, we shall say that the formation of unions and intersections are "operations" on S, just as we speak of addition and multiplication being operations on N (the set of natural numbers). Furthermore, just as the natural numbers are "closed" with respect to addition and multiplication, the set S is "closed" with respect to the formation of unions and intersections. The operations $\cup$ and $\cap$ on S (the family of subsets of X) also exhibit commutative, associative, and distributive properties which we now list. If $A, B, C \in$ S, then

$$\text{(i)} \quad A \cup B = B \cup A,$$
$$\text{(ii)} \quad A \cap B = B \cap A,$$
$$\text{(iii)} \quad A \cup (B \cup C) = (A \cup B) \cup C,$$
$$\text{(iv)} \quad A \cap (B \cap C) = (A \cap B) \cap C,$$
$$\text{(v)} \quad A \cap (B \cup C) = (A \cap B) \cup (A \cap C),$$
$$\text{(vi)} \quad A \cup (B \cap C) = (A \cup B) \cap (A \cup C).$$

Note that (i) and (ii) state that $\cup$ and $\cap$ are *commutative* operations, (iii) and (iv) state that $\cup$ and $\cap$ are *associative* operations, and (v) and

(vi) state that each of the operations $\cup$ and $\cap$ is *distributive* with respect to the other. These properties of the operations $\cup$ and $\cap$ can be *proved*—they need not be assumed—and to prove them we will need to use our theorem for the equality of sets (Section 3–2). As an example, let us prove Property (iii): if A, B, $C \in S$, then $A \cup (B \cup C) = (A \cup B) \cup C$.

Proof. We must prove two things:

(1) $A \cup (B \cup C) \subset (A \cup B) \cup C$ and (2) $(A \cup B) \cup C \subset A \cup (B \cup C)$.

If we can prove *both* (1) and (2) we will know by the theorem for the equality of sets that $A \cup (B \cup C) = (A \cup B) \cup C$. To prove (1), suppose that $x \in A \cup (B \cup C)$. Then $x \in A$ *or* $x \in B \cup C$ (by definition of a union), and we consider these two cases separately. If $x \in B \cup C$, then $x \in B$ *or* $x \in C$ and, if $x \in B$, $x \in A \cup B$. Thus if $x \in B \cup C$, we can assert that $x \in C$ *or* $x \in A \cup B$, so that $x \in (A \cup B) \cup C$. If, on the other hand, $x \in A$, then $x \in A \cup B$, so that $x \in (A \cup B) \cup C$. We have now shown that if x is any element of $A \cup (B \cup C)$, then x is necessarily an element of $(A \cup B) \cup C$ and $A \cup (B \cup C) \subset (A \cup B) \cup C$.

To prove (2), suppose that $y \in (A \cup B) \cup C$. Then $y \in A \cup B$ *or* $y \in C$, and we consider these two cases separately. If $y \in A \cup B$, then $y \in A$ *or* $y \in B$, and if $y \in B$, $y \in B \cup C$. Thus if $y \in A \cup B$, we can assert that $y \in A$ *or* $y \in B \cup C$, so that $y \in A \cup (B \cup C)$. If, on the other hand, $y \in C$, then $y \in B \cup C$ and $y \in A \cup (B \cup C)$. We have now shown that if y is *any* element of $(A \cup B) \cup C$, then y is necessarily an element of $A \cup (B \cup C)$ and $(A \cup B) \cup C \subset A \cup (B \cup C)$.

Finally, then, since each of the sets $A \cup (B \cup C)$ and $(A \cup B) \cup C$ is a subset of the other, it must be true that $A \cup (B \cup C) = (A \cup B) \cup C$. Note that Property (iii) has been proved on the tacit assumption that both sets are nonempty. It should be clear, however, that if either of the sets were empty the other would be empty also (why?) and the two sets would still be equal.

A Venn diagram may be used to illustrate the principle just proved. The set $A \cup (B \cup C)$ is the set identified in Fig. 3–5 by the *horizontal* shading. The vertical shading is applied *first* to identify $B \cup C$ and *then* the horizontal shading is applied to identify $A \cup (B \cup C)$. In Fig. 3–6, the set $(A \cup B) \cup C$ is identified by the *vertical* shading. The horizontal shading is applied *first* to identify $A \cup B$, and *then* $(A \cup B) \cup C$ is identified. The set crossed by horizontal shading in the diagram of Fig. 3–5 *appears* to be the same set as the one crossed by vertical shading in the diagram of Fig. 3–6.

In addition to constructing sets from given sets by forming unions, intersections, and differences, we can form *products*, but before we can

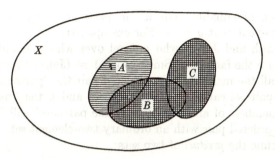

FIGURE 3–5

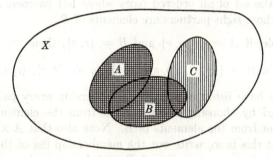

FIGURE 3–6

plunge into a discussion of products of sets we must understand the concept of *ordered pair*. We shall base our definition of *ordered pair* on the rather vague concept of "lefthandedness," since a stricter definition would present difficulties beyond the scope of this text.*

DEFINITION. An *ordered pair* is a two-element set which has one of its members designated as the *left* member.

Since an ordered pair differs from an ordinary two-element set, a special notation is needed. An ordered pair having elements a and b with a designated as the left member is written (a, b), and if b is designated as the left member, the ordered pair is written (b, a).

To say that an element is the *right* member of an ordered pair means, simply, that it is *not the left* member and to say that two ordered pairs are *equal* means that they have the same left members and, also, the same right members. We formally define equality of ordered pairs as follows.

DEFINITION. $(a, b) = (c, d)$ if, and only if, $a = c$ and $b = d$.

---

* John L. Kelley, *Introduction to Modern Algebra*, New York, Van Nostrand, 1960 (sec. 13).

This definition of equality tells us, in particular, that $(a, b) \neq (b, a)$, except in the event that $a = b$. For example, the members of a comedy team, say Smith and Jones, who quarrel over whom should receive top billing illustrate the fact that (Smith, Jones) $\neq$ (Jones, Smith).

We shall call the members of an ordered pair the "partners"; if (x, y) is any ordered pair we call x the "left partner" and y the "right partner." Calling the members of an ordered pair "the partners" will help us avoid confusing an ordered pair with an ordinary two-element set.

We now define the *product* of two sets:

DEFINITION. If A and B are any sets we define $A \times B$ (read "A cross B") to be the set of all ordered pairs whose left partners are elements of A and whose right partners are elements of B.

For example, if $A = \{p, q, r\}$ and $B = \{r, s\}$, then

$$A \times B = \{(p, r),\ (p, s),\ (q, r),\ (q, s),\ (r, r),\ (r, s)\}.$$

Note that we have listed in $A \times B$'s membership every pair that could be constructed by choosing a left partner from the elements of A and a right partner from the elements of B. Note also that $A \times B \neq B \times A$ [to prove that this is so, write out the membership list of the set $B \times A$ and note that the sets $A \times B$ and $B \times A$ have no elements (pairs) in common except the element (r, r)].

EXERCISE GROUP 3–4

1. Given that $R = \{s, t, u\}$ and $S = \{u, v\}$. Write the membership list of $R \times S$.

2. Given that $A = \{a, b\}$ and $B = \{c, d\}$. Write the membership lists of the nonempty subsets of $A \times B$.

3. If C is a set having m distinct elements and D is a set having n distinct elements, how many elements does the set $C \times D$ have? Support your answer with a verbal argument.

4. If E is any set, what can you say about the sets $E \times \emptyset$ and $\emptyset \times E$? Explain.

5. If A, B, C are any sets, *prove* that

$$A \times (B \cup C) = (A \times B) \cup (A \times C)$$

and

$$A \times (B \cap C) = (A \times B) \cap (A \times C).$$

6. Given that $H = \{p, q\}$, $K = \{r, s\}$, and $L = \{t, u, v\}$. Write out the memberships of $H \times (K \times L)$ and $(H \times K) \times L$. Are they the same? Explain.

7. Let $X = \{u, v, w\}$, $A = \{u\}$, $B = \{v\}$, $C = \{w\}$, $D = \{u, v\}$, $E = \{u, w\}$, and $F = \{v, w\}$. Then $S = \{\emptyset, A, B, C, D, E, F, X\}$ is the family of all subsets of X. Insert the correct element of S in each square of the following tables.

∪	∅	A	B	C	D	E	F	X
∅								
A								
B								
C								
D								
E								
F								
X								

∩	∅	A	B	C	D	E	F	X
∅								
A								
B								
C								
D								
E								
F								
X								

These tables remind us of a miniature arithmetic (Chapter 2), with ∅, A, B, C, D, E, F, X playing the role of "numbers" and ∪, ∩ playing the role of "operations." What special role is played by the "number" ∅ in the ∪ table and by the "number" X in the ∩ table?

CHAPTER 4

MAPPINGS AND OPERATIONS

4–1 Mapping of a set into a set. Suppose that we are given two sets, say $A = \{p, q, r\}$ and $B = \{s, t, u, v\}$, and are given the following command: "Draw a set of arrows so that *one* arrow points away from *each* element of A and points toward *some* element of B." To carry out this command it will be convenient to write the membership lists of A and B side by side, either vertically or horizontally. The result might be represented graphically as follows:

$$A = \{p, q, r\}$$

$$B = \{s, t, u, v\}$$

Whoever gave the command could just as well have said, "Illustrate a mapping of A into B by drawing arrows," and this remark leads us to the definition of the most general kind of mapping—a mapping *into:*

DEFINITION. If S and T are any sets (not necessarily distinct), then a mapping of S *into* T is a rule that associates *each* element of S with a *unique* (one and only one) element of T.

The drawing of arrows has nothing to do with mappings; arrows only serve as a crutch to help us visualize how the elements of one set are associated with the elements of another, just as Venn diagrams help us to visualize unions, intersections, and differences of sets. If a set of students enters a classroom and takes seats, the students are being mapped *into* the set of seats—no arrows are drawn; if at a dance the orchestra leader shouts "ladies' choice" and each lady present chooses a gentleman partner (not always the one she wants), the set of ladies is being mapped *into* the set of men—no arrows are actually drawn; if, after grading a set of examination papers, a teacher returns them to the class, the set of papers is being mapped *into* the set of students—no arrows are actually drawn. Whenever we associate the members of one set with those of another set we are performing a mapping. For example, when we look at a map of the state of Minnesota, we mentally associate each point of the state itself with a point on the map, and by so doing we are mapping one set of points into another.

The arrows we have drawn above to graphically illustrate a mapping of A into B could have been drawn in many different ways; we indicate a few of them here:

$$\{p,\ q,\ r\} \qquad \{p,\ q,\ r\} \qquad \{p,\ q,\ r\}$$
$$\{s,\ t,\ u,\ v\} \qquad \{s,\ t,\ u,\ v\} \qquad \{s,\ t,\ u,\ v\}$$

In each of the mappings of A into B that have so far been illustrated, *one* arrow leaves *each* element of A and arrives at *some* element of B. The question now arises as to how many ways it is possible to map A into B. To determine this we proceed as follows. Consider that in any mapping of A into B we draw an arrow from p to some *one* of the elements $s,\ t,\ u,\ v$ (four choices), we draw an arrow from q to some *one* of the elements $s,\ t,\ u,\ v$ (again, four choices), and we draw an arrow from r to some *one* of the elements $s,\ t,\ u,\ v$ (again, four choices). Thus the total number of ways that we can *choose* to draw a set of three arrows is $4 \cdot 4 \cdot 4 = 4^3 = 64$; therefore there are 64 different ways that we can elect to map A into B. We now formulate a general rule for finding the total number of ways that one nonempty set may be mapped into another:

RULE. If A has m elements and B has n elements, then A may be mapped into B in n^m different ways.

To prove this rule, consider that we are to draw m arrows (one from each element of A) and we have a choice of pointing each arrow toward any *one* of the n elements of B. Therefore, since the first arrow can be pointed in n ways, the second arrow can be pointed in n ways, ..., the mth arrow can be pointed in n ways, the set of m arrows can be pointed in $n \cdot n \cdot n \cdot \ldots \cdot n$ (m factors altogether) $= n^m$ ways.

We shall want to choose some symbols to represent mappings and we adopt letters from the Greek alphabet: α (alpha), β (beta), γ (gamma), δ (delta), and so on. The particular mapping

$$A = \{p,\ q,\ r\}$$
$$B = \{s,\ t,\ u,\ v\}$$

discussed earlier might be represented by, say, α, and we define α by using either the notation

$$\alpha\colon p \to t,\ q \to v,\ r \to u$$

or the notation

$$\alpha p = t,\ \alpha q = v,\ \alpha r = u.$$

Each of these notations defines α by describing how α associates the elements of A with the elements of B. When we see the definition of α written in either of these notations we know that α associates p with t, q with v, and r with u, and we call t the *image* of p, v the *image* of q, and u the *image* of r. The notation αp (read "alpha p") is merely a symbol for the *image* of p, αq is a symbol for the *image* of q, and αr is a symbol for the *image* of r when the rule α is used to map A into B. We are also justified in writing $\alpha p = t$, since αp and t are symbols for *the same* element.

Later we will be concerned with "equal" mappings and we now define what it means to say that two mappings are *equal*:

DEFINITION. If α and β are mappings of a set S into a set T and if $\alpha a = \beta a$ for each $a \in S$, we say that α and β are *equal*, and write $\alpha = \beta$. If $\alpha a \neq \beta a$ for at least one element $a \in S$, we write $\alpha \neq \beta$.

In short, two mappings of one set into another are equal if, and only if, the application of the mappings to any element of one set always yields the same image.

EXERCISE GROUP 4–1

1. In each of the following problems a set A, a set B, and a mapping of A into B is given. Illustrate each mapping by listing the memberships of A and B side by side and actually drawing arrows.

(a) $A = \{a_1, a_2, a_3\}$, $B = \{b_1, b_2\}$;
$\beta a_1 = b_1$, $\beta a_2 = b_1$, $\beta a_3 = b_2$.

(b) $A = \{h, j, k\}$, $B = \{p, q, r, h\}$;
$\gamma h = r$, $\gamma j = h$, $\gamma k = p$.

(c) $A = \{3, 5, 7\}$, $B = \{2, 4, 6\}$;
$\alpha: 3 \rightarrow 4$, $5 \rightarrow 6$, $7 \rightarrow 2$.

2. For each part of Problem 1 find in how many ways A may be mapped into B. For part (a) illustrate, by drawing arrows, each possible mapping of A into B.

3. Let A be the set of students sitting in the front row of your mathematics class and let

$$B = \{17, 18, 19, 20, 21, 22, 23, 24, 25\}.$$

By drawing arrows, map A into B, using the rule (mapping) which associates each member of A with his age on his *last* birthday; also map A into B using the rule which associates each member of A with his age on his *nearest* birthday. (Enlarge set B if necessary.)

4–2 Mappings of a set onto a set. We have talked about mappings of a set *into* a set, and we now go on to discuss mappings of a set *onto* a set. First, however, let us remember that a mapping of a set S *into* a set T assigns *each* element of S to *some one* of the elements of T; for example, if $S = \{a, b, c\}$ and $T = \{p, q\}$, the following illustrates a perfectly good mapping of S *into* T:

$$S = \{a, b, c\}$$

$$T = \{p, q\}$$

Note that *each* element of S is assigned to (associated with) an element of T. Note further that, with respect to the particular mapping pictured, q is the image of each element of S and p has no elements of S assigned to it! Now, there are eight (2^3) mappings of S into T, and only two of the eight mappings assign each element of S to the same element of T, i.e., the mapping given above and the mapping which assigns a, b, and c to p. In all of the other six mappings p is the image of at least one of the elements of S, as is q also. One of these six mappings is

$$S = \{a, b, c\}$$

$$T = \{p, q\}$$

If this mapping is symbolized by β, we have $\beta a = q$, $\beta b = p$, $\beta c = q$, or $\beta : a \to q$, $b \to p$, $c \to q$. β is indeed a mapping of S *into* T, but it is something more: since *each* element of T is the image of *some* element of S we call β a mapping of S *onto* T. We now formally define what we mean when we speak of a mapping of one set *onto* another:

DEFINITION. If α is a mapping of a set S into a set T, then α is called a mapping of S *onto* T if for *each* $y \in T$ there is *some* $x \in S$ such that $\alpha x = y$.

Note particularly the word "some," which means the same as "at least one." As an example of a mapping *onto*, consider a banquet hall with 40 tables, each with 4 place settings. If 157 persons enter the hall and each person takes a seat at a table, we have a mapping of the set of persons *onto* the set of tables; each person is associated with *one* table and each table is associated with *at least one* person.

Mathematicians often use the word "function" instead of "mapping." When a mathematician speaks of "a function on S to T" he is talking about a mapping of S *into* T or, what is the same thing, a mapping of S *onto* a *subset* of T; he calls S the *domain* of the function and the subset of T (the set of images) the *range* of the function. In more advanced courses in mathematics mappings are usually called "functions" whenever the domain and range are *sets of numbers*. Mathematics courses such as algebra, trigonometry, analytic geometry, and calculus deal with the study of functions and their properties.

1. If A is a set having m elements and B is a set having n elements, what can you say about the number of mappings of A *onto* B if $m < n$? if $m = n$?

2. If $A = \{t, u, v, w\}$ and $B = \{p, q, r\}$ how many of the mappings of A *into* B are mappings *onto*?

3. Make a set S of 20 different license numbers selected at random from cars on the campus parking lot and assign each license number to its last (rightmost) digit in the set $T = \{0, 1, 2, 3, 4, 5, 6, 7, 8, 9\}$. Is your mapping a mapping of S *onto* T or merely a mapping of S *into* T?

4–3 One-to-one mappings. A type of mapping of special interest to us is a $1:1$ (one-to-one) mapping of a set onto a set, which we now define:

DEFINITION. A mapping α is called a $1:1$ mapping of S onto T if, and only if, (1) α is a mapping of S *onto* T and (2) any two distinct elements of S have distinct images.

Observe that for a mapping to be a $1:1$ mapping of S onto T it must meet two requirements: Each element of T must be the image of *some* element of S and no two distinct elements of S have the same image in T. We immediately observe that a $1:1$ mapping of a finite set S onto a finite set T is possible if and only if S and T have the same number of elements. If S has *more* elements than T, then any mapping of S *onto* T must associate two or more elements of S with the same element of T and it would not be true that distinct elements of S have distinct images; on the other hand, if S has *fewer* elements than T any mapping of S *into* T would leave at least one element of T not associated with an element of S. In other words, no mapping of S *into* T could possibly be a mapping *onto*.

Assuming that it is possible to map a set S $1:1$ onto a set T, we ask how many such mappings there actually are. Before attempting to answer this question we consider a simple example using sets $A = \{a, b\}$ and $B = \{c, d\}$ and actually illustrating all possible mappings of A into B:

We see that *two* of these mappings are $1:1$, and we might now ask if there are *three* $1:1$ mappings of a set of three elements onto a set of three elements. To answer this question, we consider the following sets C and D:

$$C = \{p, q, r\},$$
$$D = \{s, t, u\}.$$

If we now try to illustrate a 1 : 1 mapping of C onto D we must draw an arrow *from p to* one of the elements s, t, u, and we see that we have *three* choices. After drawing an arrow from p we must draw an arrow *from* q but, since distinct elements of C must have distinct images, we cannot assign q to p's image; in other words, in drawing an arrow from q we have only *two* choices. Finally, we see that two of the three elements in D have been made images of p and q, and therefore there is only *one* way that we can draw an arrow from r. To summarize, there are *three* ways to draw our *first* arrow, *two* ways to draw our *second* arrow, and *one* way to draw our *third* arrow, or there are $3 \cdot 2 \cdot 1 = 6$ ways of drawing a set of three arrows to illustrate a 1 : 1 mapping of C onto D.

In general, if we have sets

$$E = \{a_1, a_2, \ldots, a_n\} \quad \text{and} \quad F = \{b_1, b_2, \ldots, b_n\},$$

both having n elements, we see that we have n ways to draw a first arrow, $n - 1$ ways to draw a second arrow, $n - 2$ ways to draw a third arrow, $\ldots$, and only one way to draw the last (nth) arrow. Thus the total number of 1 : 1 mappings of E onto F (or F onto E) is

$$n \cdot (n - 1) \cdot (n - 2) \cdot \ldots \cdot 3 \cdot 2 \cdot 1.$$

Note that this expression is the product of all the natural numbers from 1 through n. Mathematicians have a special way of writing products of this kind; they write it $n!$, read "n factorial." In our earlier example dealing with sets C and D, each with three elements, the number of 1 : 1 mappings turned out to be $3! = 1 \cdot 2 \cdot 3 = 6$.

Recall now, in the definition of a mapping *into* (Section 4–1), the phrase "not necessarily distinct," which was enclosed in parentheses. The significance of this phrase is that a set may be mapped *into* itself. For example, if a teacher has a stack of examination papers on his desk he may decide to rearrange the papers, say in alphabetical order. When he finishes doing this he has exactly the same stack (set) of papers; he has merely mapped the set into itself (in this case he has actually mapped the set 1 : 1 *onto* itself). A 1 : 1 mapping of any finite set onto itself does nothing but rearrange the members of the set; mathematicians call such mappings *permutations*. If we are asked how many permutations there are of a set of 5 elements, we are merely being asked in how many different orders the membership list can be written or in how many ways we can map the set 1 : 1 onto itself. The answer, of course, is $1 \cdot 2 \cdot 3 \cdot 4 \cdot 5$ or $5!$.

It is evident by now that if we know of a 1 : 1 mapping of a set S onto a set T, we also know of a 1 : 1 mapping of T onto S, the mapping that assigns each image in T to the element in S that was assigned to it originally. The second mapping is called the *inverse* of the first. As an example, consider the sets $S = \{a, b, c\}$ and $T = \{d, e, f\}$ with the $1 : 1$ mapping $\alpha : a \rightarrow e, \ b \rightarrow f, \ c \rightarrow d$. Under the mapping α the elements $d, \ e, \ f$ are the images, and the mapping which maps each one of these images onto the element of which it is the image is called the *inverse of* α, or α-*inverse*, written α^{-1}. We describe this inverse by writing $\alpha^{-1} : e \rightarrow a, f \rightarrow b, d \rightarrow c$.

In terms of a set of arrows, these $1 : 1$ mappings may be illustrated by writing

$$\alpha : \quad \begin{array}{c} \{a, b, c\} \\ \diagdown\!\!\diagup\diagdown \\ \{d, e, f\} \end{array}$$

and

$$\alpha^{-1} : \quad \begin{array}{c} \{a, b, c\} \\ \diagup\diagdown\!\!\diagup \\ \{d, e, f\} \end{array}$$

In the language of arrows, then, it appears that the inverse of any $1 : 1$ mapping is the $1 : 1$ mapping that merely reverses the direction of the arrows, and either of the two mappings is the *inverse* of the other. Thus, in the example just illustrated, we have $(\alpha^{-1})^{-1} = \alpha$; that is, the inverse of alpha-inverse is equal to (the same as) alpha. (Recall the definition of equal mapping given in Section 4–1.)

We shall now discuss the *composition of mappings*, particularly the composition of $1 : 1$ mappings. To illustrate what is meant by this concept let us examine the sets

$$R = \{a, b, c\}, \quad S = \{p, q, r\}, \quad T = \{s, t, u\},$$

together with the mappings

$$\alpha : a \rightarrow r, b \rightarrow q, c \rightarrow p \quad \text{and} \quad \beta : p \rightarrow t, q \rightarrow s, r \rightarrow u.$$

We first set up the following tables:

$$
\begin{array}{lcl}
\alpha a = r & & \beta p = t \\
\alpha b = q & \text{and} & \beta q = s \\
\alpha c = p & & \beta r = u
\end{array}
$$

In the table on the right we can substitute αc for p, αb for q, and αa for r, to obtain the following table:

$$
\begin{array}{l}
\beta \alpha c = t \\
\beta \alpha b = s \\
\beta \alpha a = u
\end{array}
$$

Observe that we actually have a mapping, $\beta\alpha$, of R onto T that is described by either the above table or the notation $\beta\alpha : a \rightarrow u,\ b \rightarrow s,$
$c \rightarrow t$; $\beta\alpha$ is the mapping obtained by "composing" the mappings α
and β. One can perhaps best describe the mapping $\beta\alpha$ by saying what it
accomplishes: $\beta\alpha$ maps the elements of R onto S and then maps the
images in S onto T. In other words, $\beta\alpha$ is the mapping α applied to the
elements of R *followed by* the mapping β applied to the images $\alpha a,\ \alpha b,$
αc. In the diagram below α and β are illustrated by the short arrows
and $\beta\alpha$ by the long arrows.

$$R \qquad S \qquad T$$

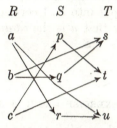

Note in this example that $\beta\alpha$ is a $1 : 1$ mapping of R onto T and that $\beta\alpha$
has an inverse which we symbolize by $(\beta\alpha)^{-1}$. Specifically,

$$(\beta\alpha)^{-1} : s \rightarrow b,\ t \rightarrow c,\ u \rightarrow a.$$

It can also be seen in this example that $(\beta\alpha)^{-1}$ accomplishes the same
thing as mapping T onto S with the mapping β^{-1} *followed by* the mapping
of the images $\beta^{-1}s = q$, $\beta^{-1}t = p$, and $\beta^{-1}u = r$ onto R by α^{-1}. In
other words, we may say that $(\beta\alpha)^{-1} = \alpha^{-1}\beta^{-1}$. These mappings
indeed are *the same*, since they accomplish the same result when applied
to the elements of the set T.

When one set is mapped $1 : 1$ onto another we often say that we have
put the sets into "one-to-one correspondence." To be more precise, when
we put a set A into $1 : 1$ correspondence with a set B, we pair each element of A with an element of B in such a way that distinct elements of A
are paired with distinct elements of B. When an element of A is paired
with an element of B, that element of B is automatically paired with the
element of A, so that, in effect, a $1 : 1$ mapping of A onto B and the
inverse $1 : 1$ mapping of B onto A have both been accomplished. Talking
about $1 : 1$ correspondences, then, is just another way of talking about
$1 : 1$ mappings.

It is convenient to use $1 : 1$ correspondence terminology when talking
about *counting*. What do we really do when we *count* a finite set? To
answer this question by saying "we find the number of elements in the
set" really does not answer the question at all, for to find the number

of elements in the set we must first count the set. When we count any finite set S we try to find a set in the unending sequence of sets

$$\{1\}, \{1, 2\}, \{1, 2, 3\}, \{1, 2, 3, 4\}, \{1, 2, 3, 4, 5\}, \ldots$$

with which the set S can be put into 1 : 1 correspondence. If, for example, set S can be put into 1 : 1 correspondence with the set $\{1, 2, 3, 4, 5, 6, 7\}$, we say that S has 7 elements and, in general, we say that a set S has n elements if, and only if, S can be put into 1 : 1 correspondence with the set $\{1, 2, 3, 4, \ldots, n\}$. We call the sets $\{1\}, \{1, 2\}, \{1, 2, 3\}, \ldots$ *cardinal sets*, and if a set S can be put into 1 : 1 correspondence with the cardinal set $\{1, 2, 3, \ldots, n\}$ we say that n is the *cardinal number* of S; in short, n is the number of elements in S.

EXERCISE GROUP 4–3

1. Let sets A, B, C each have cardinal number 4. Illustrate by *drawing arrows* (a) a 1 : 1 mapping, α, of A into B, (b) a 1 : 1 mapping, β, of B onto C, (c) the mapping $\beta\alpha$ of A onto C, (d) the mapping α^{-1}, (e) the mapping β^{-1}, (f) the mapping $(\beta\alpha)^{-1}$, and (g) the mapping $\alpha^{-1}\beta^{-1}$.

2. In Problem 1, how many 1:1 mappings are there of A onto B? of B onto C? How many *different* mappings of A onto C can be formed by composing a 1:1 mapping of B onto C with a 1:1 mapping of A onto B?

3. Let $A = \{x, y, z\}$ and let α, β, γ be the mappings of A onto A defined by

$$\alpha : x \rightarrow y, \; y \rightarrow z, \; z \rightarrow x,$$
$$\beta : x \rightarrow z, \; y \rightarrow x, \; z \rightarrow y,$$
$$\gamma : x \rightarrow y, \; y \rightarrow x, \; z \rightarrow z.$$

Find $\alpha\gamma : x \rightarrow \underline{\quad}, \; y \rightarrow \underline{\quad}, \; z \rightarrow \underline{\quad},$
$$\gamma\beta : x \rightarrow \underline{\quad}, \; y \rightarrow \underline{\quad}, \; z \rightarrow \underline{\quad},$$
$$\gamma\beta\alpha : x \rightarrow \underline{\quad}, \; y \rightarrow \underline{\quad}, \; z \rightarrow \underline{\quad},$$
$$\alpha\beta : x \rightarrow \underline{\quad}, \; y \rightarrow \underline{\quad}, \; z \rightarrow \underline{\quad}.$$

4. Consider the set of mappings $M = \{\alpha, \beta, \gamma, \delta\}$ of the set $B = \{x, y\}$ into B. These mappings are defined as follows:

$$\alpha : x \rightarrow x, y \rightarrow y; \; \beta : x \rightarrow y, y \rightarrow x; \; \gamma : x \rightarrow x, y \rightarrow x; \; \delta : x \rightarrow y, y \rightarrow y.$$

In the squares of the table below insert the mappings (elements of M) obtained by composing the mappings at the top with (followed by) the mappings at the left. For example, $\beta\alpha = \beta$.

	α	β	γ	δ
α				
β	β			
γ				
δ				

What can you say about the set M with respect to the composition of its elements?

5. Given $A = \{1, 2, 3, 4, 5\}$ and $B = \{a, b, c, d\}$. Illustrate, by drawing arrows, a mapping of A into B which is *not* onto, and a mapping of A into B which *is* onto.

6. Given the set $D = \{1, 2, 3, 4\}$ and the following 1:1 mappings of D onto D:

$$\alpha : 1 \rightarrow 4,\ 2 \rightarrow 3,\ 3 \rightarrow 2,\ 4 \rightarrow 1,$$

$$\beta : 1 \rightarrow 2,\ 2 \rightarrow 3,\ 3 \rightarrow 4,\ 4 \rightarrow 1,$$

$$\gamma : 1 \rightarrow 2,\ 2 \rightarrow 4,\ 3 \rightarrow 1,\ 4 \rightarrow 3.$$

Find each of the following compositions:

$$\alpha\beta : 1 \rightarrow \underline{\hspace{1cm}},\ 2 \rightarrow \underline{\hspace{1cm}},\ 3 \rightarrow \underline{\hspace{1cm}},\ 4 \rightarrow \underline{\hspace{1cm}},$$

$$\alpha^{-1}\beta\gamma^{-1} : 1 \rightarrow \underline{\hspace{1cm}},\ 2 \rightarrow \underline{\hspace{1cm}},\ 3 \rightarrow \underline{\hspace{1cm}},\ 4 \rightarrow \underline{\hspace{1cm}},$$

$$(\alpha\beta)^{-1}\gamma : 1 \rightarrow \underline{\hspace{1cm}},\ 2 \rightarrow \underline{\hspace{1cm}},\ 3 \rightarrow \underline{\hspace{1cm}},\ 4 \rightarrow \underline{\hspace{1cm}}.$$

7. Let α be a 1:1 mapping of R onto S and let β be a 1:1 mapping of S onto T. Prove that $\beta\alpha$ is a 1:1 mapping of R onto T. [*Hint:* Show (a) that $\beta\alpha$ is a mapping of R *into* T, (b) that $\beta\alpha$ is a mapping of R *onto* T, and (c) that distinct elements of R have distinct images.]

8. Let

$$R = \{a, b, c\},\quad S = \{p, q, r\},\quad T = \{u, v, w\}.$$

If

$$\alpha : a \rightarrow q,\ b \rightarrow r,\ c \rightarrow p \quad \text{and} \quad (\beta\alpha)^{-1} : u \rightarrow c,\ v \rightarrow a,\ w \rightarrow b,$$

where β is a 1:1 mapping of S onto T, then

$$\beta : p \rightarrow \underline{\hspace{1cm}},\ q \rightarrow \underline{\hspace{1cm}},\ r \rightarrow \underline{\hspace{1cm}}.$$

9. Let α and β be mappings of the set N of natural numbers into itself, defined as follows: for each $x \in N$, $\alpha x = x^2$ and $\beta x = 2 \cdot x + 3$. Find the mappings $\alpha\alpha$, $\beta\beta$, $\alpha(\beta\alpha)$, and $(\alpha\beta)\alpha$.

10. Using the mappings α and β of Problem 7, prove that $(\beta\alpha)^{-1} = \alpha^{-1}\beta^{-1}$. [*Hint:* Show that these mappings produce the same image when applied to any element of T.]

4–4 Operations on a set. In mathematics we are constantly performing "operations." For example, when we write $3 \cdot 5 = 15$ we say that we have performed an operation, and when we write $5 + 2 = 7$ we also say that we have performed an operation. In the first case we performed a multiplication and in the second case an addition, and we call these "operations on the set N of natural numbers." We now propose to answer the question "What, exactly, is meant by *an operation on a set?*" We state the following definition:

DEFINITION. An *operation* on a nonempty set S is a mapping of $S \times S$ into S.

Recalling that a mapping of one set into another is a rule that associates each element of the one set with a unique element of the other, we see that an "operation on S" is a rule that associates each ordered pair of elements of S with an element of S. Thus multiplication is an operation on the set N, for it associates each ordered pair of natural numbers with a natural number. In the example above, $3 \cdot 5 = 15$ is merely our customary way of saying that the ordered pair $(3, 5)$ is mapped into (associated with) the natural number 15.

As an example of an abstract operation on a set S, suppose that $S = \{p, q, r\}$ and α is a mapping of $S \times S$ into S defined by

$$\alpha : (p, p) \to q, (p, q) \to p, (p, r) \to p, (q, q) \to r, (q, r) \to q,$$
$$(q, p) \to q, (r, r) \to p, (r, p) \to r, (r, q) \to r.$$

By drawing arrows, we can illustrate the mapping α as follows:

$$S \times S = \{(p, p), (p, q), (p, r), (q, p), (q, q), (q, r), (r, p), (r, q), (r, r)\}$$

$$S = \{ \qquad p, \qquad q, \qquad r \qquad \}$$

The mapping α of $S \times S$ into S is indeed "an operation on S" somewhat awkwardly described, but an operation nevertheless.

To avoid such awkward descriptions of operations as the one above we "invent" a symbol of operation. Suppose we use the symbol $*$ for our example and write

$$p * p = q, p * q = p, p * r = p, q * q = r, q * r = q,$$
$$q * p = q, r * r = p, r * p = r, r * q = r.$$

It should be understood that

$$p * p, p * q, p * r, \ldots, r * q$$

are merely symbols for the images of the ordered pairs (p, p), (p, q), $(p, r), \ldots, (r, q)$ under the mapping α, and $p * p = q$, $p * q = p, \ldots,$ and so on, *by definition of* α. We can describe the mapping α still more neatly by setting up an operation table, just as we would a multiplication table or an addition table. Thus, for the present example, we would have the following table:

(Right)

(Left)

*	p	q	r
p	q	p	p
q	q	r	q
r	r	r	p

If we wish to find at a glance to which element of S the mapping α assigns an element (an ordered pair) of $S \times S$, we look for the left partner in the left column and for the right partner in the top row. Then the element of S in the left partner's row and in the right partner's column is the desired image. Thus, for example, the table tells us that $q * r = q$, so that we immediately know $\alpha : (q, r) \to q$. It will now be of interest to find the total number of operations on S (mappings of $S \times S$ into S). Since there are 9 elements in $S \times S$ and each one can be assigned an image in 3 ways, the total number of ways that we can map $S \times S$ into S is

$$3 \cdot 3 \cdot 3 \cdot 3 \cdot 3 \cdot 3 \cdot 3 \cdot 3 \cdot 3 = 3^9 = 19{,}683.$$

The particular operation * was chosen for illustrative purposes by the "eenie, meenie, miney, moe" process from 19,683 different possible operations on S. This example should illustrate the complete generality we are trying to achieve; p, q, r represent arbitrary elements of set S and * represents an arbitrarily selected operation on S. Nothing that has been said or done should imply that p, q, r are "numbers" or that * bears any resemblance to familiar operations such as addition or multiplication.

4–5 Mathematical systems. From this point on we shall be greatly concerned with what we call "mathematical systems" and we shall study a number of mathematical systems to discover what properties they have in common and in what ways they differ. To begin with, we should forget any notion that a mathematical system is necessarily a number system (whatever that is). When, in the previous section, we talked about a set $S = \{p, q, r\}$ with an operation * defined on S, we were, in fact, talking about a mathematical system; the elements p, q, r of S *were not* *numbers.* We shall now give a precise definition of the concept of mathematical system:

DEFINITION. A *mathematical system* is any nonempty set S together with one or more operations defined on S.

If S is a nonempty set and $*$ is an operation on S, then the set S together with the operation $*$ is a mathematical system, and we use the symbol $\{S; *\}$ for this system. The set of natural numbers together with the operation $+$ is the mathematical system designated by $\{N; +\}$, the set of natural numbers together with the operation $\cdot$ is the mathematical system designated by $\{N; \cdot\}$, and the set of natural numbers together with the operations $+$ and $\cdot$ is the mathematical system designated by $\{N; +, \cdot\}$. It should be noted that $-$ (minus) is *not* an operation on N, for $-$ does *not* map each ordered pair of natural numbers into a natural number. For example, if $-$ were an operation on N, the ordered pair $(3, 5)$ would be associated with a natural number; in other words, $3 - 5$ would be a natural number. Similar remarks apply to $\div$, so that $\{N; -\}$ and $\{N; \div\}$ are meaningless symbols. We already have defined subtraction and division on the natural numbers, but we did *not* claim that $-$ and $\div$ were *operations* on N.

We warned earlier that a mathematical system is not necessarily a number system; however, a number system *is* a special kind of mathematical system. Before defining a "number system" and discussing its properties, we should be reminded that a set may have more than one operation defined on it. If S is a nonempty set with distinct operations $*$ and $\#$ (sharp) defined on S, we will use the notation $\{S; *, \#\}$ to designate the set S *together with* the operations $*$ and $\#$. If S is a set with two elements, then $S \times S$ has four elements and there are $2^4 = 16$ mappings of $S \times S$ into S; hence there are sixteen possible operations on S. If $*, \#, \$, \ldots, \triangle$ represent the sixteen operations, we would use the symbol $\{S; *, \#, \$, \ldots, \triangle\}$ to designate the set S *together with* its sixteen operations.

With the above remarks in mind we now concentrate on the definitions of *commutativity, associativity,* and *distributivity* with respect to operations on a set.

DEFINITION. (1) If $*$ is any operation on a set S and if $a * b = b * a$ whenever $a, b \in S$, we say that $*$ is a *commutative* operation; (2) if $*$ is any operation on S and if $(a * b) * c = a * (b * c)$ whenever $a, b, c \in S$, we say that $*$ is an *associative* operation; and (3) if $*$ and $\#$ are any operations on S and if $a \# (b * c) = (a \# b) * (a \# c)$ whenever $a, b, c \in S$, we say that $\#$ is *distributive* "over" $*$ or that it is distributive "with respect to" $*$.

After reviewing the properties of the natural numbers in Section 2–2 we see that we have a set N with operations $+$ and $\cdot$ defined on N, where (1) both $+$ and $\cdot$ are commutative, (2) both $+$ and $\cdot$ are associative, and (3) $\cdot$ is distributive over $+$. With this example in mind, we now formally state the definition of *number system*:

DEFINITION. The mathematical system $\{S; *, \#\}$ is a number system if, and only if, the operations $*$ and $\#$ are both commutative, both associative, and one of them is distributive over the other.

We immediately see that we have every right to call the mathematical system $\{N; +, \cdot\}$ a "number system" and to call the symbols 0, 1, 2, 3, ... "numbers." In fact, whenever a system $\{S; *, \#\}$ is a number system we call the elements of S numbers. As an illustration, let us consider an abstract number system $\{R; \oplus, \odot\}$, where $R = \{p, q\}$ and the operations $\oplus$ and $\odot$ are defined by the following tables:

$\oplus$	p	q
p	p	q
q	q	p

$\odot$	p	q
p	p	p
q	p	q

We have asserted that $\{R; \oplus, \odot\}$ is a number system, but now we must *prove* it. Unfortunately, there is no easy way to do this; we must grind out the proof by the "strong-arm" method. For example, to show that $\oplus$ is commutative we must prove the truth of the following *statement:* If $a, b \in R$, then $a \oplus b = b \oplus a$. In this statement, a and b are both *variables*, and R is the *domain* of each. To prove that $\oplus$ is commutative we must show that $a \oplus b = b \oplus a$ remains true when a and b are allowed to vary independently over their domain; in other words, we must show that $a \oplus b = b \oplus a$ if $a = p$ and $b = p$, if $a = p$ and $b = q$, if $a = q$ and $b = p$, and, finally, if $a = q$ and $b = q$. This amounts, then, to trying all possible cases, which we now do:

if $a = p, b = p$, then $a \oplus b = p \oplus p = p$ and $b \oplus a = p \oplus p = p$, so that $a \oplus b = b \oplus a$;

if $a = p, b = q$, then $a \oplus b = p \oplus q = q$ and $b \oplus a = q \oplus p = q$, so that $a \oplus b = b \oplus a$;

if $a = q, b = p$, then $a \oplus b = q \oplus p = q$ and $b \oplus a = p \oplus q = q$, so that $a \oplus b = b \oplus a$;

if $a = q, b = q$, then $a \oplus b = q \oplus q = p$ and $b \oplus a = q \oplus q = p$, so that $a \oplus b = b \oplus a$.

Having let the variables a and b assume all possible "values" in their domain, we see that $a \oplus b = b \oplus a$ whenever $a, b \in R$, and so $\oplus$ is commutative. Without going into so much detail we observe that

$$p \oplus (p \oplus p) = p \oplus p = p \quad \text{and} \quad (p \oplus p) \oplus p = p \oplus p = p,$$
$$p \oplus (p \oplus q) = p \oplus q = q \quad \text{and} \quad (p \oplus p) \oplus q = p \oplus q = q,$$
$$p \oplus (q \oplus p) = p \oplus q = q \quad \text{and} \quad (p \oplus q) \oplus p = q \oplus p = q,$$
$$p \oplus (q \oplus q) = p \oplus p = p \quad \text{and} \quad (p \oplus q) \oplus q = q \oplus q = p,$$
$$q \oplus (p \oplus p) = q \oplus p = q \quad \text{and} \quad (q \oplus p) \oplus p = q \oplus p = q,$$
$$q \oplus (p \oplus q) = q \oplus q = p \quad \text{and} \quad (q \oplus p) \oplus q = q \oplus q = p,$$
$$q \oplus (q \oplus p) = q \oplus q = p \quad \text{and} \quad (q \oplus q) \oplus p = p \oplus p = p,$$
$$q \oplus (q \oplus q) = q \oplus p = q \quad \text{and} \quad (q \oplus q) \oplus q = p \oplus q = q,$$

and therefore $\oplus$ is associative. We have now seen what had to be done in order to prove that $\oplus$ is both commutative and associative. It is left as an exercise to show in a similar way that $\odot$ is both commutative and associative. Finally, we prove that $\odot$ is distributive over $\oplus$ by letting a, b, c vary independently over their domain R and showing that $a \odot (b \oplus c) = (a \odot b) \oplus (a \odot c)$ is true in every possible case:

$$p \odot (p \oplus p) = p \odot p = p \quad \text{and} \quad (p \odot p) \oplus (p \odot p) = p \oplus p = p,$$
$$p \odot (p \oplus q) = p \odot q = p \quad \text{and} \quad (p \odot p) \oplus (p \odot q) = p \oplus p = p,$$
$$p \odot (q \oplus p) = p \odot q = p \quad \text{and} \quad (p \odot q) \oplus (p \odot p) = p \oplus p = p,$$
$$p \odot (q \oplus q) = p \odot p = p \quad \text{and} \quad (p \odot q) \oplus (p \odot q) = p \oplus p = p,$$
$$q \odot (p \oplus p) = q \odot p = p \quad \text{and} \quad (q \odot p) \oplus (q \odot p) = p \oplus p = p,$$
$$q \odot (p \oplus q) = q \odot q = q \quad \text{and} \quad (q \odot p) \oplus (q \odot q) = p \oplus q = q,$$
$$q \odot (q \oplus p) = q \odot q = q \quad \text{and} \quad (q \odot q) \oplus (q \odot p) = q \oplus p = q,$$
$$q \odot (q \oplus q) = q \odot p = p \quad \text{and} \quad (q \odot q) \oplus (q \odot q) = q \oplus q = p,$$

and therefore $\odot$ is distributive over $\oplus$.

We have shown that $\{R; \oplus, \odot\}$ is a number system, since both operations are commutative, both are associative, and one is distributive over the other. Recall that there are sixteen possible operations on the set $R = \{p, q\}$. We have proved only that R is a number system with respect to the operations $\oplus$ and $\odot$ defined by the tables. With respect to some other pair of operations the set R may not be a number system, if one or more of the three essential requirements fails to hold.

In any number system $\{S; *, \#\}$ either $*$ is distributive over $\#$, $\#$ is distributive over $*$, or each operation is distributive over the other. In a number system where one operation is distributive over the other, it is common practice to call the distributive operation "multiplication" and the other "addition," not because these operations bear any resemblance to ordinary multiplication and addition but because the names sound familiar and the elements (numbers) behave with respect to these operations just as natural numbers behave with respect to ordinary multiplication and addition. In the example just discussed we elect to call $\odot$ "multiplication" and $\oplus$ "addition," since $\odot$ is distributive over $\oplus$, just as multiplication is distributive over addition in the natural number system. Hereafter we will often use the symbols $\oplus$ and $\odot$ for the operations of a number system and whenever we say "consider the number system $\{S; \oplus, \odot\}$," we imply that $\odot$ is distributive over $\oplus$.

Suppose that $\{S; *\}$ is a mathematical system. If there is an element $e \in S$ such that $e * a = a$ and $a * e = a$ for all $a \in S$, we call e an "identity element with respect to $*$." It is necessary to specify "with respect to $*$," because e may *not* be an identity element with respect to some other operation on S. As an example, we have the system $\{N; +\}$ in which 0 is an identity element with respect to $+$, since $0 + n = n + 0 = n$ for all $n \in N$. Similarly, 1 is an identity element with respect to $\cdot$ in the system $\{N; \cdot\}$, since $1 \cdot n = n \cdot 1 = n$ for all $n \in N$. We now prove a useful fact about an identity element:

THEOREM. If e is an identity element of the system $\{S; *\}$, then e is unique; in other words, there is no other identity element with respect to $*$.

To prove this theorem, we assume that f is also an identity element of $\{S; *\}$. Then, since f is an identity element, $f * e = e * f = e$ and, since e is an identity element, $e * f = f * e = f$. Therefore, in the statement $f * e = e * f$ we may substitute f for $f * e$ and e for $e * f$ to obtain $f = e$. We have proved that $\{S; *\}$ has no identity element different from e.

We have said that 0 is *an* identity element of $\{N; +\}$, but we now know that it is proper to say that 0 is *the* identity element of $\{N; +\}$. Similarly, we may also call 1 *the* identity element of $\{N; \cdot\}$. In some number systems, the system may have an identity element with respect to each operation. For example, in the number system $\{R; \oplus, \odot\}$ used as an example, we see that $p \oplus p = p$ and $p \oplus q = q \oplus p = q$, so that $p \oplus a = a \oplus p = a$ for all $a \in R$ and, furthermore, we see that $q \odot p = p \odot q = p$ and $q \odot q = q$, so that $q \odot a = a \odot q = a$ for all $a \in R$. Hence p is the identity element of R with respect to $\oplus$ and q is the identity element of R with respect to $\odot$. Since we have called $\oplus$ "addition" and $\odot$ "multiplication," we call p the *additive* identity of R, and q the *multiplicative* identity of R.

1. Consider Problem 7 of Exercise Group 3–4, where $X = \{u, v, w\}$, $\mathcal{S}$ is the family of all subsets of X, and the system $\{\mathcal{S}; \cup, \cap\}$ is a number system. Does $\mathcal{S}$ have identity elements with respect to the operations $\cup$ and $\cap$? If so, what are they?

2. There are sixteen operations on the set $P = \{r, s\}$. Make a table defining *each one*. Identify those operations which are commutative and identify *two* operations which are *not* associative. Identify those operations which are commutative and with respect to which there is an identity element.

3. Consider Problem 4 of Exercise Group 4–3, where $M = \{\alpha, \beta, \gamma, \delta\}$ is the set of all mappings of the set B into itself. The table defines an operation (the composition of mappings) on M. Is there an identity element with respect to this operation? If so, what is it? Show that the composition of mappings is *not* commutative. Is there more than one identity element?

4. Consider the mathematical system $\{S; *\}$, where $S = \{x, y, z, w\}$ and $*$ is an operation on S defined by the following table:

*	x	y	z	w
x	x	w	x	z
y	y	w	y	x
z	x	y	z	w
w	z	y	w	x

(a) The operation $*$ is one of how many possible operations on S?

(b) Is there an identity element in S and, if there is, what is it?

(c) How do you know that $*$ is *not* commutative?

(d) To prove $*$ associative it would be necessary to show that

$$a * (b * c) = (a * b) * c$$

for *all* $a, b, c \in S$. The *truth* of the statement $a * (b * c) = (a * b) * c$ would have to be verified in how many different cases?

(e) To prove that $*$ is *not* associative it would be necessary to show that $a * (b * c) \neq (a * b) * c$ in how many different cases? Do it!

CHAPTER 5

GROUPS

5–1 Definition of a group. Examples. A *group* is a special kind of mathematical system encountered so frequently that we single it out for special study. We now formally define a group:

DEFINITION. A mathematical system $\{S; *\}$ is called a *group* if, and only if, it possesses the following three properties:

  (i) $*$ is an associative operation,
 (ii) there exists a unique element† $e \in S$, called the identity element, such that $e * a = a * e = a$ for all $a \in S$, and
(iii) for each $a \in S$ there exists an element $b \in S$, called an inverse of a, such that $b * a = a * b = e$.

Let us now examine some mathematical systems with which we are already familiar and determine whether or not they are groups. Is the system $\{N; +\}$ a group? To answer this question we ask ourselves if (i) addition is associative, if (ii) there is an identity element with respect to addition, and if (iii) each natural number has an inverse. We know that addition of natural numbers is associative and we know that the natural number 0 is the (unique) additive identity, since $0 + n = n + 0 = n$ for every $n \in N$. However, not every natural number has an inverse; for example, there is no natural number which added to 5 will yield the identity 0. Therefore, since one of the three necessary conditions fails to hold, $\{N; +\}$ is *not* a group.

Next, we ask if the system $\{R; \oplus\}$ studied in Section 4–5 is a group. We proved that $\oplus$ is an associative operation and we know that $p \oplus p = p$ and $p \oplus q = q \oplus p = q$, and therefore the element $p \in R$ is the identity element. Finally, since $p \oplus p = p$ (the identity) and $q \oplus q = p$ (the identity), we see that p and q each have an inverse. Therefore, the system $\{R; \oplus\}$ is a group.

As another example of a group, consider the system $\{S; \oplus\}$, where $S = \{0, 1, 2\}$ and $\oplus$ is the operation on S defined by Table 5–1 (see next page). This may, at first glance, look like a strange kind of arithmetic, but it is commonly used and extremely practical. In Table 5–1 we are adding as we would on a clock, in this case a three-hour clock (Fig. 5–1) instead of a twelve-hour clock (e.g., on an ordinary clock, 11 o'clock + 4 hours = 3 o'clock).

---

† The reader should review the theorem on the uniqueness of identity elements, Section 4–5.

TABLE 5-1

⊕	0	1	2
0	0	1	2
1	1	2	0
2	2	0	1

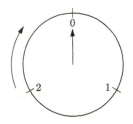

FIGURE 5-1

In the above table, if a, $b \in S$, then $a \oplus b$ is always the element $c \in S$ which is obtained by starting at a on the clock and rotating the hand b spaces clockwise. Thus $2 \oplus 2$ means the element $1 \in S$ obtained when we start the hand at 2 and rotate it 2 spaces clockwise, and so we write $2 \oplus 2 = 1$. First we note that $0 \in S$ is the identity element; for *any* $a \in S$ we have $0 \oplus a = a \oplus 0 = a$. We note, also, that each element $a \in S$ has an inverse: $0 \oplus 0 = 0$ and $1 \oplus 2 = 2 \oplus 1 = 0$, so that, therefore, 0 is an inverse of 0, 2 is an inverse of 1, and 1 is an inverse of 2. To prove that $\oplus$ is associative it will be necessary to show that $a \oplus (b \oplus c) = (a \oplus b) \oplus c$ when a, b, c are allowed to vary independently over their domain $S = \{0, 1, 2\}$. In other words, it will be necessary to try twenty-seven cases (do it!). A simpler way of proving that $\oplus$ is an associative operation will be studied later.

5–2 Inherent properties of a group. In Section 2–2 we discussed properties which we *assumed* the natural numbers to possess and we *proved* that the natural numbers possessed additional properties. We shall now prove that a group possesses properties other than the three which it possesses by definition. As we list each of these properties we will be talking about a mathematical system $\{S; *\}$ which we assume to be a group. The first property concerns the uniqueness of inverses:

(1) Each element of a group has a *unique* inverse.

To prove (1), suppose that $a \in S$. Since $\{S; *\}$ is a group, there is an element $b \in S$ (an inverse of a) such that $b * a = a * b = e$. If we suppose that an element $c \in S$ is also an inverse of a, it is true that $c * a = a * c = e$. Then, starting with the fact that $a * b = e$, we may argue as follows:

$$c * (a * b) = c * e = c, \qquad \text{(Why?)}$$
$$c * (a * b) = (c * a) * b, \qquad \text{(Why?)}$$
$$(c * a) * b = c, \qquad \text{(Why?)}$$
$$c * a = e, \qquad \text{(Why?)}$$
$$e * b = c, \qquad \text{(Why?)}$$
$$b = c. \qquad \text{(Why?)}$$

Thus our conclusion is that the element a has no inverse distinct from b; the element a has one *and only one* inverse. We are now justified in speaking of *the* inverse of an element, and we shall adopt the notation a^{-1} for the unique inverse of an element $a \in S$. Having proved (1) we are in a position to show that a group possesses further properties:

(2) If a, b, $c \in S$ and $a * b = a * c$, then $b = c$. This is known as the *left cancellation property*.

(3) If a, b, $c \in S$ and $b * a = c * a$, then $b = c$. This is known as the *right cancellation property*.

(4) If a, $b \in S$, there are unique elements x, $y \in S$ such that $a * x = b$ and $y * a = b$.

(5) If $a \in S$, then the inverse of a^{-1} is a; that is, $(a^{-1})^{-1} = a$.

(6) If a, $b \in S$, then $(a * b)^{-1} = b^{-1} * a^{-1}$.

To prove (2) we use the fact that there is a unique element $a^{-1} \in S$ such that $a^{-1} * a = e$ for each $a \in S$, and we write our argument in steps:

$$a * b = a * c, \qquad \text{(Given)}$$
$$a^{-1} * (a * b) = a^{-1} * (a * c), \qquad \text{(Why?)}$$
$$(a^{-1} * a) * b = (a^{-1} * a) * c, \qquad \text{(Why?)}$$
$$e * b = e * c, \qquad \text{(Why?)}$$
$$b = c. \qquad \text{(Why?)}$$

(Note that the substitution principle has been used freely.) The proof of Property (3) is similar and is left to the reader.

To prove (4) we write $a * x = b$ as *an equation to be solved*. We argue that *if* $x \in S$ and *if* $a * x = b$, then

$$a^{-1} * (a * x) = a^{-1} * b, \qquad \text{(Why?)}$$
$$(a^{-1} * a) * x = a^{-1} * b, \qquad \text{(Why?)}$$
$$e * x = a^{-1} * b, \qquad \text{(Why?)}$$
$$x = a^{-1} * b. \qquad \text{(Why?)}$$

We now substitute to see if $a^{-1} * b$ is a value of $x \in S$ such that $a * x = b$ is true. We see that it is for

$$a * (a^{-1} * b) = (a * a^{-1}) * b, \qquad \text{(Why?)}$$
$$(a * a^{-1}) * b = e * b, \qquad \text{(Why?)}$$
$$e * b = b. \qquad \text{(Why?)}$$

Therefore

$$a * (a^{-1} * b) = b. \qquad \text{(Why?)}$$

We have shown that *if* there is an element $x \in S$ such that $a * x = b$, then x *must be* the element $a^{-1} * b$ and no other. Furthermore, we have shown that *if* $x = a^{-1} * b$, then $a * x = b$. In short, we have proved the existence of a unique element $x \in S$ such that $a * x = b$. It is left for the reader to prove that there is a unique element $y \in S$ such that $y * a = b$.

The proof of Property (5) is left to the reader. Finally, to prove (6) we note that $a * b \in S$ has a unique inverse, $(a * b)^{-1}$, such that $(a * b) * (a * b)^{-1} = e$ for all $a, b \in S$. We also note that

$$
\begin{aligned}
(a * b) * (b^{-1} * a^{-1}) &= a * [b * (b^{-1} * a^{-1})] &&\text{(Why?)} \\
&= a * [(b * b^{-1}) * a^{-1}] &&\text{(Why?)} \\
&= a * [e * a^{-1}] &&\text{(Why?)} \\
&= a * a^{-1} &&\text{(Why?)} \\
&= e.
\end{aligned}
$$

But, by Property (4), there exists a *unique* element $x \in S$ such that $(a * b) * x = e$, so that $(a * b)^{-1} = b^{-1} * a^{-1}$. We commonly state this property by saying: "In a group whose operation is $*$, the inverse of the $*$ of any two elements is equal to the $*$ of their inverses in reverse order."

EXERCISE GROUP 5-1

1. Prove Property (3), the right cancellation property.

2. Complete the proof of Property (4); that is, show that there exists a *unique* element $y \in S$ such that $y * a = b$.

3. Let $S = \{p, q, r\}$ and let $M = \{\alpha, \beta, \gamma, \delta, \epsilon, \zeta\}$ (alpha, beta, gamma, delta, epsilon, zeta) be the set of all 1:1 mappings of S onto S. Show that the mathematical system $\{M; *\}$ is a group, where $*$ is the composition of mappings. The following steps are suggested:

    (a) Illustrate these mappings by drawing arrows and let each of the symbols $\alpha, \beta, \gamma, \delta, \epsilon, \zeta$ represent one of the mappings.

    (b) Let θ (theta), ψ (psi), ω (omega) be variables each having domain M, and show that $\theta * \psi \in M$ for all $\theta, \psi \in M$. (Here you will need to make a table to show that $*$ actually is an operation on M.)

    (c) Find the identity mapping by consulting the table you have made.

    (d) Prove that $*$ is an associative operation on M by showing that

$$\theta * (\psi * \omega) = (\theta * \psi) * \omega$$

for *all* $\theta, \psi, \omega \in M$. (You will *not* need to prove all 216 cases! Simply take any element $x \in S$ and show that both $\theta * (\psi * \omega)$ and $(\theta * \psi) * \omega$ applied to x will yield the same image.)

4. Let S be any nonempty set with a finite but unspecified number of elements and let M be the set of all 1:1 mappings of S onto S. If $*$ represents the composition of mappings, show that $\{M; *\}$ is a group. Here you will not be able to construct a table. Refer to Problem 7 of Exercise Group 4–3 for a hint. If $\theta, \psi \in M$, what can you conclude about $(\theta * \psi)^{-1}$ and $\psi^{-1} * \theta^{-1}$? Why?

5. Explain why it was necessary to prove both cancellation properties. Property (4) tells us that there are unique elements $x, y \in S$ such that $a * x = b$ and $y * a = b$. Is it true that $y = x$?

6. Prove Property (5).

5–3 Permutation groups. In Section 4–3 it was mentioned that a 1:1 mapping of a finite set onto itself is called a *permutation*. Permutations are of interest to us because certain sets of permutations give us beautiful examples of groups. Problems 3 and 4 at the end of Section 5–2 illustrate that one can always find an example of a group by merely using the set M of all permutations of some finite set and letting the composition of permutations be the operation on M.

As an example of a group, let us consider the set of all "transformations" of an equilateral triangle. The concept of a transformation will be explained shortly, and we will see that it is in reality a permutation. Let us imagine that we have a piece of cardboard cut in the shape of an equilateral triangle and that, because we are going to be turning our cardboard triangle over, we have labeled the vertices A, B, and C on both sides. Let us now imagine that we place our triangle on a table and trace around it with chalk or pencil (Fig. 5–2). We now have an image of our cardboard triangle drawn on the table and coinciding with the cardboard

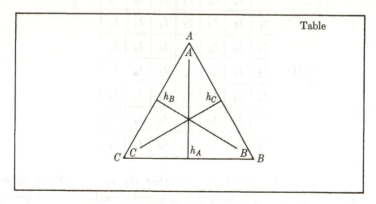

FIGURE 5–2

triangle itself. If the triangle is now moved in such a way that it returns to coincidence with its image, the motion will be called a *transformation*. There are exactly six such transformations—no more and no less. Let us designate these by t_1, t_2, t_3, t_4, t_5, t_6, where t_1 is the transformation that does nothing to the triangle (the identity transformation), t_2 rotates the triangle 120° clockwise (240° counterclockwise), t_3 rotates the triangle 240° clockwise (120° counterclockwise), t_4 reflects the triangle about the altitude h_A, t_5 reflects the triangle about the altitude h_B, and t_6 reflects the triangle about the altitude h_C. It is now evident that these transformations really do nothing but map the set $\{A, B, C\}$ of vertices 1:1 onto itself:

$$t_1: \quad A \rightarrow A, \quad B \rightarrow B, \quad C \rightarrow C;$$
$$t_2: \quad A \rightarrow B, \quad B \rightarrow C, \quad C \rightarrow A;$$
$$t_3: \quad A \rightarrow C, \quad B \rightarrow A, \quad C \rightarrow B;$$
$$t_4: \quad A \rightarrow A, \quad B \rightarrow C, \quad C \rightarrow B;$$
$$t_5: \quad A \rightarrow C, \quad B \rightarrow B, \quad C \rightarrow A;$$
$$t_6: \quad A \rightarrow B, \quad B \rightarrow A, \quad C \rightarrow C.$$

We shall now see what happens when we compose these transformations, that is, when we apply any two of them in succession. Let i, j denote variables each having the set $\{1, 2, 3, 4, 5, 6\}$ as its domain, and for any i, j let $t_i * t_j$ denote the transformation obtained when t_j is followed by t_i. If we let i and j vary independently over their domain, we obtain the results shown in the table

(Right)

	$*$	t_1	t_2	t_3	t_4	t_5	t_6
	t_1	t_1	t_2	t_3	t_4	t_5	t_6
	t_2	t_2	t_3	t_1	t_6	t_4	t_5
(Left)	t_3	t_3	t_1	t_2	t_5	t_6	t_4
	t_4	t_4	t_5	t_6	t_1	t_2	t_3
	t_5	t_5	t_6	t_4	t_3	t_1	t_2
	t_6	t_6	t_4	t_5	t_2	t_3	t_1

It is a fairly simple matter to show that the system $\{T; *\}$, where T designates the set $\{t_1, t_2, t_3, t_4, t_5, t_6\}$ of transformations of an equilateral triangle, is a group. First of all, we must show that $*$ is an associative

operation; in other words, that $t_i * (t_j * t_k) = (t_i * t_j) * t_k$ for all
$i, j, k \in \{1, 2, 3, 4, 5, 6\}$. We note that for any i, j, k, $t_i * (t_j * t_k)$ is the
transformation obtained when the application of $t_j * t_k$ is followed by
the application of t_i. But since $t_j * t_k$ is the transformation obtained when
the application of t_k is followed by the application of t_j, $t_i * (t_j * t_k)$ is
the transformation obtained if the transformations t_k, t_j, t_i, in that order,
are applied to the triangle. Similarly, we note that $(t_i * t_j) * t_k$ is the
transformation obtained when the application of t_k is followed by the
application of $t_i * t_j$. But since $t_i * t_j$ is the transformation obtained when
the application of t_j is followed by the application of t_i, $(t_i * t_j) * t_k$ is the
transformation obtained if the transformations t_k, t_j, t_i, in that order, are
applied to the triangle. Therefore we may say that

$$t_i * (t_j * t_k) = (t_i * t_j) * t_k$$

for all $i, j, k \in \{1, 2, 3, 4, 5, 6\}$. This proves that $*$ is an associative
operation.

Since $*$ is a *binary* operation on T (it maps the ordered *pairs* of $T \times T$
into T), the expression $t_i * t_j * t_k$ has no meaning unless it is defined.
We shall define $t_i * t_j * t_k$ to mean $t_i * (t_j * t_k)$ and hereafter, if we wish
to do so, we may omit the parentheses when writing $t_i * (t_j * t_k)$; we
may also omit parentheses when writing $(t_i * t_j) * t_k$ (why?).

Next, inspection of the table tells us that

$$t_1 * t_i = t_i * t_1 = t_i$$

for all

$$i \in \{1, 2, 3, 4, 5, 6\},$$

and therefore t_1 is the unique identity element in the set T. Finally,
each element in T has an inverse, for, by examination of the table, we see
that

$$t_1 * t_1 = t_1, \quad t_2 * t_3 = t_3 * t_2 = t_1, \quad t_4 * t_4 = t_1, \quad t_5 * t_5 = t_1, \quad t_6 * t_6 = t_1.$$

Therefore, $\{T; *\}$ is a group, as we wished to show.

The group $\{T; *\}$ is often called "the group of symmetries of an equilateral triangle." A "symmetry" of a geometric figure is nothing more
than a motion that leaves the *appearance* of the figure unchanged. Were
it not for the fact that we have the vertices of our equilateral triangle
labeled, the application of any one of the symmetries $t_1, t_2, t_3, t_4, t_5, t_6$
would leave the appearance of the triangle unchanged. Note that the six
symmetries (transformations) of the equilateral triangle fall into two
classes: the *rotations* t_1, t_2, t_3 and the *reflections* t_4, t_5, t_6. We now note that
the system $\{R; *\}$ of rotations, where $R = \{t_1, t_2, t_3\}$, is also a group.

The table is as follows:

*	t_1	t_2	t_3
t_1	t_1	t_2	t_3
t_2	t_2	t_3	t_1
t_3	t_3	t_1	t_2

We know that $*$ is an associative operation on R because $*$ is an associative operation on T, of which R is a subset, that t_1 is the identity element, and that each element of $\{R; *\}$ has an inverse. The group $\{R; *\}$ is an example of what we call a *subgroup*. We note from the symmetry of the table that the operation $*$ is commutative. If a diagonal is drawn from the upper left-hand corner of the table to the lower right-hand corner, squares which are symmetrically located with respect to this diagonal contain the same element. In other words, in all cases the square found in t_i's column and t_j's row contains the same element as the square found in t_j's column and t_i's row. This is not the case in the $\{T; *\}$ table. For example, $t_5 * t_4 = t_3$ and $t_4 * t_5 = t_2$; i.e., the element found in t_4's column and t_5's row is t_3, while the element found in t_5's column and t_4's row is t_2. In other words, the operation $*$ on T is *not* commutative. In general, if the table for an operation on a finite set is available, it is possible to tell whether or not the operation is commutative; if the table is symmetric with respect to the downward diagonal through the table the operation is commutative, otherwise, it is not.

In connection with the group $\{T; *\}$ it was mentioned earlier that the set T is the set of all 1:1 mappings of the set $\{A, B, C\}$ of vertices onto itself, and we adopted the symbols $t_1, t_2, t_3, t_4, t_5, t_6$ for these mappings. We could also use the following symbols for these mappings:

$$\begin{pmatrix} A & B & C \\ A & B & C \end{pmatrix} = t_1, \quad \begin{pmatrix} A & B & C \\ B & C & A \end{pmatrix} = t_2, \quad \begin{pmatrix} A & B & C \\ C & A & B \end{pmatrix} = t_3,$$

$$\begin{pmatrix} A & B & C \\ A & C & B \end{pmatrix} = t_4, \quad \begin{pmatrix} A & B & C \\ C & B & A \end{pmatrix} = t_5, \quad \begin{pmatrix} A & B & C \\ B & A & C \end{pmatrix} = t_6.$$

With these symbols it is an easy matter to compose the mappings without the aid of a cardboard triangle or arrows. First we note that we can arrange the letters in the top row of any mapping symbol in any manner, so long as we make the corresponding rearrangement in the bottom row; for example,

$$\begin{pmatrix} A & B & C \\ C & B & A \end{pmatrix} = \begin{pmatrix} C & A & B \\ A & C & B \end{pmatrix},$$

since both symbols indicate the mapping $A \to C, B \to B, C \to A$. Using this obvious principle, let us find the mapping (permutation) obtained when t_6 is composed with (followed by) t_2. We write

$$t_2 * t_6 = \begin{pmatrix} A & B & C \\ B & C & A \end{pmatrix} * \begin{pmatrix} A & B & C \\ B & A & C \end{pmatrix} = \begin{pmatrix} B & A & C \\ C & B & A \end{pmatrix} * \begin{pmatrix} A & B & C \\ B & A & C \end{pmatrix}.$$

Note that *both* rows of the second (left) permutation have been rearranged so that the letters in the top row are arranged in the same order as the letters in the bottom row of the first (right) permutation. We can now "cancel" the BAC and write

$$\begin{pmatrix} \cancel{B} & \cancel{A} & \cancel{C} \\ C & B & A \end{pmatrix} * \begin{pmatrix} A & B & C \\ \cancel{B} & \cancel{A} & \cancel{C} \end{pmatrix} = \begin{pmatrix} A & B & C \\ C & B & A \end{pmatrix} = t_5,$$

so that $t_2 * t_6 = t_5$. It is easily seen that what we have done is equivalent to drawing arrows, where we have simply replaced the diagram

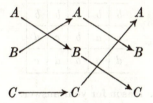

with the diagram

$$A \longrightarrow B \longrightarrow C$$
$$B \longrightarrow A \longrightarrow B$$
$$C \longrightarrow C \longrightarrow A$$

We have been dealing with the group of permutations on the vertices A, B, C of an equilateral triangle. Let us now consider the set of all permutations on *any* set of three distinct elements. Since the elements of the set are unspecified, we may adopt for them the symbols $1, 2, 3$. Remember, $1, 2, 3$ are *not* the elements of the set but merely *symbols* or *labels* for the elements, whatever they are. There are six ways to map this set onto itself:

$$\begin{pmatrix} 1 & 2 & 3 \\ 1 & 2 & 3 \end{pmatrix}, \begin{pmatrix} 1 & 2 & 3 \\ 2 & 3 & 1 \end{pmatrix}, \begin{pmatrix} 1 & 2 & 3 \\ 3 & 1 & 2 \end{pmatrix}, \begin{pmatrix} 1 & 2 & 3 \\ 1 & 3 & 2 \end{pmatrix}, \begin{pmatrix} 1 & 2 & 3 \\ 3 & 2 & 1 \end{pmatrix}, \begin{pmatrix} 1 & 2 & 3 \\ 2 & 1 & 3 \end{pmatrix}.$$

These are called *the permutations on a set of three elements* and they may be composed in the manner used in the previous example. For example, omitting a symbol of operation, we would write

$$\begin{pmatrix} 1 & 2 & 3 \\ 1 & 3 & 2 \end{pmatrix}\begin{pmatrix} 1 & 2 & 3 \\ 3 & 2 & 1 \end{pmatrix} = \begin{pmatrix} \cancel{3} & \cancel{2} & \cancel{1} \\ 2 & 3 & 1 \end{pmatrix}\begin{pmatrix} 1 & 2 & 3 \\ \cancel{3} & \cancel{2} & \cancel{1} \end{pmatrix} = \begin{pmatrix} 1 & 2 & 3 \\ 2 & 3 & 1 \end{pmatrix}.$$

EXERCISE GROUP 5-2

1. Consider the group $\{T; *\}$ of transformations of an equilateral triangle. Given that $U = \{t_1, t_4\}$, $V = \{t_1, t_5\}$, $W = \{t_1, t_6\}$, construct a table for each of the systems $\{U; *\}$, $\{V; *\}$, $\{W; *\}$ and show that each is a subgroup of $\{T; *\}$.

2. Given $S = \{a, b, c, d\}$ and the associative operation $\#$ on S defined by the following table:

#	a	b	c	d
a	b	b	a	b
b	b	b	b	b
c	a	b	c	d
d	a	b	d	c

Is $\{S; \#\}$ a group? Give a reason for your answer.

3. Fill in the following table for an operation $*$ on the set $\{p, q, r, s\}$ in such a way that q is an identity element and each element has an inverse, but in an otherwise arbitrary manner.

*	p	q	r	s
p				
q				
r				
s				

How many cases would you have to check to prove that the operation $*$ as you have defined it is associative? Check the truth of the following statements:

$$(p * p) * r = p * (p * r),$$
$$(p * r) * s = p * (r * s),$$
$$(s * p) * r = s * (p * r).$$

Have you or have you not proved whether $*$ as you have defined it is associative? Give a reason for your answer.

4. Let $P = \{p_1, p_2, p_3, p_4, p_5, p_6\}$ be the set of permutations on a set of three elements, where

$$p_1 = \begin{pmatrix} 1 & 2 & 3 \\ 1 & 2 & 3 \end{pmatrix}, \quad p_2 = \begin{pmatrix} 1 & 2 & 3 \\ 2 & 3 & 1 \end{pmatrix},$$

and so on. Letting the composition of permutations be the operation on P, construct the table for this operation. Enter the proper p_i in each square of the table, using only the symbols

$$\begin{pmatrix} 1 & 2 & 3 \\ 1 & 2 & 3 \end{pmatrix}, \quad \begin{pmatrix} 1 & 2 & 3 \\ 2 & 3 & 1 \end{pmatrix}, \dots$$

to obtain your answers.

5. Draw a square and label the vertices A, B, C, D. Let S be the set of symmetries (transformations) of the square. Using the composition of symmetries as the operation on S, construct the table for this operation. (Let $s_1, s_2, s_3, \dots$ represent the symmetries.) Prove that S is a group with respect to this operation. Write the subsets of S which are rotations and reflections. Is either of these a subgroup of S with respect to the group operation?

6. Perform the following compositions of permutations:

(a) $\left[\begin{pmatrix} 1 & 2 & 3 \\ 3 & 2 & 1 \end{pmatrix} \begin{pmatrix} 1 & 2 & 3 \\ 2 & 3 & 1 \end{pmatrix} \right] \begin{pmatrix} 1 & 2 & 3 \\ 1 & 3 & 2 \end{pmatrix}$

(b) $\begin{pmatrix} 1 & 2 & 3 \\ 3 & 2 & 1 \end{pmatrix} \left[\begin{pmatrix} 1 & 2 & 3 \\ 2 & 3 & 1 \end{pmatrix} \begin{pmatrix} 1 & 2 & 3 \\ 1 & 3 & 2 \end{pmatrix} \right]$

(c) $\begin{pmatrix} 1 & 2 & 3 & 4 \\ 3 & 4 & 1 & 2 \end{pmatrix} \begin{pmatrix} 1 & 2 & 3 & 4 \\ 4 & 2 & 3 & 1 \end{pmatrix}$

5–4 Isomorphisms. You may have noticed that some of the mathematical systems we have studied are vaguely similar; not similar in the sense that the sets of the systems are the same and not similar in the sense that the operations on the sets are the same, but similar in the sense that the systems *behave* in a strikingly similar way. We describe this similarity of behavior by saying that the systems are *isomorphic*, or by saying that there is an *isomorphism* between the systems. We may fairly start the explanation of this new concept by saying that there is really nothing new about it at all; an isomorphism is nothing more than a special kind of 1:1 mapping of one mathematical system onto another. We formally define an isomorphism as follows:

DEFINITION. Let $\{S; *\}$ and $\{T; \#\}$ be mathematical systems and let α be a 1:1 mapping of S onto T. Then α is called an *isomorphism* of $\{S; *\}$ onto $\{T; \#\}$ if, and only if,

$$\alpha(a * b) = (\alpha a) \# (\alpha b) \quad \text{for all} \quad a, b \in S,$$

and we say that the systems are *isomorphic*.

According to the above definition, if α is an isomorphism, and any two elements of S are selected, we will always obtain the same element in T whether we take the image of their * or the # of their images. To clarify this remark, consider the following illustration. Let

$$S = \{p, q, r, \ldots\},$$

$$\downarrow \quad \downarrow \quad \downarrow$$

$$T = \{u, v, w, \ldots\},$$

and let the mapping illustrated by the arrows be a $1:1$ mapping, α, of S onto T. Since * is an operation on S we will be able to search the membership list of S and find the element $p * q$; since # is an operation on T we will be able to search the membership list of T and find the element $u \# v$; and, if α is an isomorphism, we will also find that

$$\alpha : p * q \to u \# v,$$

or, in our other notation,

$$\alpha(p * q) = (\alpha p) \# (\alpha q).$$

The important thing about an isomorphism is that, in the image system, it *preserves certain properties* possessed by the parent system. For example: (1) If in the system $\{S; *\}$ the operation * is commutative, then the operation # is also commutative in the system $\{T; \#\}$. (2) If * is associative in the system $\{S; *\}$, then # is associative in $\{T; \#\}$. (3) If e is the identity element in the system $\{S; *\}$, then αe is the identity element in the system $\{T; \#\}$. (4) If a^{-1} is the inverse of the element a in the system $\{S; *\}$, then αa^{-1} is the inverse of αa in the system $\{T; \#\}$. We shall prove only (2); that is, if α is an isomorphism of $\{S; *\}$ onto $\{T; \#\}$ and * is associative, then # is associative also.

Proof. Our assumptions are that * is associative on S and that α is an isomorphism of $\{S; *\}$ onto $\{T; \#\}$. Letting x, y, z be *any* elements in T, we must show that $x \# (y \# z) = (x \# y) \# z$. Now since α is a mapping of S *onto* T, there are elements a, b, $c \in S$ such that $\alpha a = x$, $\alpha b = y$, $\alpha c = z$, and, since * is associative, $a * (b * c) = (a * b) * c$ and $\alpha[a * (b * c)] = \alpha[(a * b) * c]$. Finally, since α is an isomorphism,

$$\alpha[a * (b * c)] = (\alpha a) \# \alpha(b * c)$$
$$= \alpha a \# [(\alpha b) \# (\alpha c)] = x \# (y \# z),$$

and

$$\alpha[(a * b) * c] = \alpha(a * b) \# (\alpha c)$$
$$= [(\alpha a) \# (\alpha b)] \# (\alpha c) = (x \# y) \# z.$$

Thus, by the substitution principle,

$$x \# (y \# z) = (x \# y) \# z,$$

as we wished to show. The proofs of (1), (3), and (4) are left as an exercise.

We now have an immediate application for proposition (2), which we have just proved. We refer to the example of "clock arithmetic" discussed in Section 5–1. The particular system discussed there was $\{S; \oplus\}$, where $S = \{0, 1, 2\}$ and $\oplus$ was "addition on the clock." We shall now prove by a simple method that $\oplus$ is an associative operation. To this end, let us consider the system $\{R; *\}$ discussed in Section 5–3, where

$$R = \{t_1, t_2, t_3\}$$

and $*$ is the composition of rotations. Let us put the tables for $\{R; *\}$ and $\{S; \oplus\}$ side by side and consider the 1:1 mapping $t_1 \to 0$, $t_2 \to 1$, $t_3 \to 2$ of R onto S:

$*$	t_1	t_2	t_3
t_1	t_1	t_2	t_3
t_2	t_2	t_3	t_1
t_3	t_3	t_1	t_2

$\oplus$	0	1	2
0	0	1	2
1	1	2	0
2	2	0	1

We note that any element in the $\{S; \oplus\}$ table is the image, under this mapping, of the element occupying the corresponding square in the $\{R; *\}$ table. In other words, the systems $\{R; *\}$ and $\{S; \oplus\}$ are isomorphic. Since $*$ (the composition of rotations) is an associative operation on R, we know by proposition (2) that $\oplus$ is an associative operation on S.

The reader may ask why the particular mapping $t_1 \to 0$, $t_2 \to 1$, $t_3 \to 2$ was selected when there were five other 1:1 mappings of R onto S to choose from. To answer fairly, it must be confessed that we knew in advance that the particular mapping chosen was an isomorphism. However, the important thing is that we were able to exhibit a single 1:1 correspondence between R and S such that the result obtained by operating with two elements in one system will always correspond to the result obtained by operating with the corresponding elements in the other system. The existence of just *one* isomorphism makes the systems *iso-*

morphic whether or not other isomorphisms exist! We now close our brief discussion of isomorphisms. Just as we found ready use for this concept in the above example, we shall find other uses for it in later chapters.

EXERCISE GROUP 5-3

1. Let P be the set of permutations on a set of three elements with the composition of permutations as the operation on P, and let S be the set of symmetries of an equilateral triangle with the composition of symmetries as the operation on S.

   (a) Prove that these two systems are isomorphic.
   (b) By finding two symmetries whose composition is *not* commutative, prove that the composition of permutations is not commutative.
   (c) Prove that the system P is a group. Use the table constructed for Problem 4 of Exercise Group 5-2.

2. If systems $\{S; *\}$ and $\{T; \#\}$ are isomorphic and $\{S; *\}$ is a group, prove that $\{T; \#\}$ is also a group. You may use any previous results obtained.

3. Prove propositions (1), (3) and (4) of Section 5-4.

4. Let $\{S; \oplus, \odot\}$ be a number system and suppose that α is an isomorphism of $\{S; \oplus\}$ onto $\{T; \#\}$ and also of $\{S; \odot\}$ onto $\{T; \$\}$. Prove that $\{T; \#, \$\}$ is a number system. You must prove that $\#$, $\$$ are both commutative and associative, and that $\$$ is distributive over $\#$. Make use of results previously obtained.

5. We were able to observe earlier in this section that the 1:1 mapping $t_1 \rightarrow 0$, $t_2 \rightarrow 1$, $t_3 \rightarrow 2$ of R onto S was an isomorphism of $\{R; *\}$ onto $\{S; \oplus\}$ by noting the correspondence between elements occupying corresponding squares of the tables. Show that the $1:1$ mapping $t_1 \rightarrow 0$, $t_2 \rightarrow 2$, $t_3 \rightarrow 1$ of R onto S is also an isomorphism of $\{R; *\}$ onto $\{S; \oplus\}$.

6. Review the discussion of the *E-O* and *R-G* tables in Section 2-5. Let $A = \{E, O\}$ and $B = \{R, G\}$, and let α be the 1:1 mapping of A onto B defined by $\alpha: E \rightarrow R$, $O \rightarrow G$. Prove that α is an isomorphism of $\{A; +\}$ onto $\{B, +\}$ by actually showing that

$$\alpha(x + y) = (\alpha x) + (\alpha y) \quad \text{for } all \quad x, y \in A,$$

and similarly prove that α is an isomorphism of $\{A; \times\}$ onto $\{B; \times\}$. Prove that $\{A; +, \times\}$ is a number system.

CHAPTER 6

RELATIONS AND PARTITIONS

6–1 Relations on a set. We are all familiar with the word *relation* as it is used in everyday language, and its mathematical meaning is closely allied with ordinary usage. For example, if we know that one person "is a cousin of" another we say that "is a cousin of" is a "relation" between the persons; if one person "is taller than" another we say that "is taller than" is a "relation" between the two persons; if one brand of coffee "is stronger than" another we say that "is stronger than" is a "relation" between the two brands of coffee; if one person "is a neighbor of" another we say that "is a neighbor of" is a "relation" between the two persons. Note that although we have given examples of relations, we have not *defined* the word "relation." Our first task, then, is to arrive at a suitable definition.

It should be evident from the examples above that a relation involves two or more objects which share some common attribute (e.g., being a person or being a brand of coffee); in other words, a relation involves two or more objects which *share membership in some set*. To illustrate, we consider the relation "is stronger than," mentioned above. This relation is a relation with respect to or *on the set* of all brands of coffee. The relations "is a cousin of," "is taller than," and "is a neighbor of" are relations with respect to *sets* of persons. To be sure, the sets of persons in our examples are quite indefinite, but we can be certain that each relation is a relation with respect to *some* set.

Let us now consider a perfectly definite set: the set J whose elements are the members of the Jones family who live down the street. Below we tabulate some pertinent information about the Jones family. For our

THE JONES FAMILY

Name		Age	Height	Symbol
Albert (father)		38	5'10''	a
Barbara (mother)		35	5' 8''	b
Claudia		15	5' 7''	c
Eunice	(children)	14	5' 6''	e
Peter (twins)		13	5' 5''	p_1
Paul (twins)		13	5' 5''	p_2
Theodore		10	4'10''	t

own convenience we have assigned a symbol to represent each member of J. We may consider a number of relations which exist among the members of this family. For example, let us examine the relation $\mathcal{R}_1$, which we define to mean "is a parent of," and let us consider the *statement form* $x\mathcal{R}_1y$. Now, x and y are *variables* whose domain is the set J and if for x and y we substitute "values" from J, we obtain *statements* such as $c\mathcal{R}_1t$, $b\mathcal{R}_1b$, or $b\mathcal{R}_1e$, the first two of which are false. Of course, we are here interested in substitutions for x and y that will make the statement form $x\mathcal{R}_1y$ a true statement.

To do this in a systematic way we take the set $J \times J$ and select that subset of $J \times J$ each of whose elements (ordered pairs) has a left partner and a right partner which when substituted for x and y, respectively, in the statement form $x\mathcal{R}_1y$, will make it a *true* statement. We note that this subset of $J \times J$ is

$$K = \{(a, c), (a, e), (a, p_1), (a, p_2), (a, t), (b, c), (b, e), (b, p_1), (b, p_2), (b, t)\}.$$

The important fact to note here is that the *relation* $\mathcal{R}_1$ determines (defines) the subset K of $J \times J$. Imagine, now, that the tables are turned and that, instead of being given the relation $\mathcal{R}_1$, we are given the subset K of $J \times J$. Does the set K determine a relation on the set J? The answer is immediately an emphatic "yes," for we can scan the membership of K and see that those pairs, and only those pairs, belong to K whose left partner "is a parent of" the right partner. If (u, v) is a particular element of $J \times J$, the statement $u\mathcal{R}_1v$ is true if, and only if, $(u, v) \in K$. In other words, the statements $u\mathcal{R}_1v$ and $(u, v) \in K$ are always *both true* or *both false!* We say that $u\mathcal{R}_1v$ and $(u, v) \in K$ are *equivalent* statements.

Up to this point we have not defined what is meant by "a relation on a set" but, using the observations we have made, we are now ready to do so.

DEFINITION. A *relation* on a set S is a subset of $S \times S$.

This definition of relation is certainly a straightforward one, and you may ask how such a simple concept can be important. However, the concept and its applications will become meaningful in later chapters, when we use this notion to construct new number systems.

Since a relation on any set S is a subset of $S \times S$, and since every set is a subset of itself, the set $J \times J$ (J = the Jones family) must be a relation on J. What relation is it? Nothing more nor less than the relation "belongs to the same family as." By this we are *not* implying that by taking any arbitrary subset (relation) of $J \times J$ it will be easy to describe the relation verbally. To illustrate this point, consider the set $S = \{0, 1, 2, 3\}$ of natural numbers. Since, by our definition, a relation on S is a subset of $S \times S$, we will first want to list the membership of $S \times S$:

$$S \times S = \{(0, 0),\ (0, 1),\ (0, 2),\ (0, 3),$$
$$(1, 0),\ (1, 1),\ (1, 2),\ (1, 3),$$
$$(2, 0),\ (2, 1),\ (2, 2),\ (2, 3),$$
$$(3, 0),\ (3, 1),\ (3, 2),\ (3, 3)\}.$$

The set $S \times S$, of course, has a great many subsets (relations), 2^{16} of them to be exact, and we do not intend to list them all. We shall consider only four subsets:

$$\mathfrak{R}_1 = \{(0, 1),\ (1, 1),\ (2, 1),\ (3, 2)\},$$
$$\mathfrak{R}_2 = \{(0, 3),\ (1, 2),\ (2, 1),\ (3, 0)\},$$
$$\mathfrak{R}_3 = \{(0, 0),\ (1, 1),\ (2, 2),\ (3, 3)\},$$
$$\mathfrak{R}_4 = \{(0, 1),\ (1, 2),\ (2, 3),\ (1, 0),\ (2, 1),\ (3, 2)\}.$$

Some of these relations are more interesting to us than others because they have special features which make them capable of *brief and meaningful verbal description.* For example, we cannot describe the relation $\mathfrak{R}_1$ verbally except by actually stating the membership of the subset $\mathfrak{R}_1$, but we can describe $\mathfrak{R}_2$ verbally by saying "$\mathfrak{R}_2$ is the relation on S such that, for $a, b \in S$, $(a, b) \in \mathfrak{R}_2$ if, and only if, $a + b = 3$," or "$\mathfrak{R}_2$ is that relation on S such that, for $a, b \in S$, $a\mathfrak{R}_2 b$ if, and only if, $a + b = 3$," or, finally, "$\mathfrak{R}_2$ is the relation of *having the sum* 3." The relations $\mathfrak{R}_3$ and $\mathfrak{R}_4$ are also capable of easy verbal description. We can say "$\mathfrak{R}_3$ is that relation on S such that, for $a, b \in S$, $a\mathfrak{R}_3 b$ if, and only if, $a = b$," or "$\mathfrak{R}_3$ is the relation *is equal to,*" and we can say "$\mathfrak{R}_4$ is that relation on S such that, for $a, b \in S$, $a\mathfrak{R}_4 b$ if, and only if, $a = b + 1$ or $a + 1 = b$," or "$\mathfrak{R}_4$ is the relation *differs by* 1 *from.*"

If $\mathfrak{R}$ is *any* relation on S, the statement form $a\mathfrak{R}b$ may be read "a is in the relation $\mathfrak{R}$ to b," where $a\mathfrak{R}b$ and $(a, b) \in \mathfrak{R}$ are *equivalent* statement forms. Indeed, we have come to the point where we may use a relation symbol interchangeably in two equivalent ways. For example, with reference to the set $S = \{0, 1, 2, 3\}$, $\mathfrak{R}$ might be the relation "is two less than," and in this case we may also write $\mathfrak{R} = \{(0, 2),\ (1, 3)\}$, for, by definition of a relation, *a relation is a set,* and a pair belonging to this set is equivalent to its left partner "being two less than" its right partner.

You may have guessed by now that the relations on sets which will most concern us are those which may be described verbally (mathematically or otherwise). This is true for the most part, and the exercises which follow are designed to give practice in describing relations and in writing their membership lists if a verbal description is given.

1. Let $S = \{0, 1, 2, 3\}$ and let $\mathcal{R}$ be a relation on S.

    (a) Write the membership list of $\mathcal{R}$ if, for $a, b \in S$, $a\mathcal{R}b$ if and only if $a > b$.

    (b) Write the membership list of $\mathcal{R}$ if, for $a, b \in S$, $a\mathcal{R}b$ if and only if $a \cdot b = 2$.

    (c) Given $\mathcal{R} = \{(0, 2), (1, 3)\}$. Complete the following description: "$\mathcal{R}$ is that relation on S such that, for $a, b \in S$, $a\mathcal{R}b$ if and only if _____."

    (d) Let $\mathcal{R}_2$ and $\mathcal{R}_4$ be the relations on S mentioned above. Given that $\mathcal{R} = \mathcal{R}_2 \cap \mathcal{R}_4$, complete the following description: "$\mathcal{R}$ is that relation on S such that, for $a, b \in S$, $a\mathcal{R}b$ if, and only if, _____ and _____."

2. Let S be a convenient subset of the set of students in your mathematics class and represent each member of S by a symbol. What subsets of $S \times S$ are defined by the relations "has the same birth month as," "sits in the same row as," and "is older than"?

3. If $J =$ the Jones family, what subsets of $J \times J$ are defined by "is the same age as" and "is the same height as"? Describe verbally in more than one way the relation on J defined by the subset

$$\left\{ \begin{matrix} (a, b), \ (a, c), \ (a, e), \ (a, p_1), \ (a, p_2), \ (a, t), \ (b, c), \ (b, e), \ (b, p_1), \ (b, p_2), \\ (b, t), \ (c, e), \ (c, p_1), \ (c, p_2), \ (c, t), \ (e, p_1), \ (e, p_2), \ (e, t), \ (p_1, t), \ (p_2, t) \end{matrix} \right\}.$$

4. Let us consider the set $T = \{a, b, c, d\}$ and a mapping, α, of T into itself defined by, say,

$$\alpha : a \to b, \ b \to c, \ c \to a, \ d \to b.$$

Let us also consider the subset

$$\mathcal{R} = \{(a, b), \ (b, c), \ (c, a), \ (d, b)\}$$

of $T \times T$. By definition of a relation we see that $\mathcal{R}$ is a relation on T. If we search for a verbal description of this relation we might say that $\mathcal{R}$ means "is assigned to," "is associated with," or "is mapped into." In fact, this example illustrates that a mapping of a set into itself is a special kind of relation on the set, and we are led to an alternative definition for a mapping into: *A mapping of any nonempty set S into itself is a relation $\mathcal{R}$ on S such that each member of S appears as the left partner of one and only one member of $\mathcal{R}$.*

What mappings of the above set T into itself are the relations

$$\{(a, c), \ (b, a), \ (c, c), \ (d, a)\} \quad \text{and} \quad \{(a, a), \ (b, a), \ (c, d), \ (d, d)\}?$$

6–2 Properties of relations. Some of the relations discussed in the previous section were interesting because they could be described verbally. However, the relations that will really interest us from now on are those that possess one or more of three special properties: *reflexivity, symmetry,* and *transitivity.* We now formally define these properties.

DEFINITION. Let S be a nonempty set and let $\mathfrak{R}$ be a relation on S. Then
   (i) $\mathfrak{R}$ is said to be *reflexive* if, and only if, $(x, x) \in \mathfrak{R}$ for *every* $x \in S$,
   (ii) $\mathfrak{R}$ is said to be *symmetric* if, and only if, for all $x, y \in S$ such that $(x, y) \in \mathfrak{R}$, $(y, x) \in \mathfrak{R}$ also, and
   (iii) $\mathfrak{R}$ is said to be *transitive* if, and only if, for all $x, y, z \in S$ such that $(x, y), (y, z) \in \mathfrak{R}$, $(x, z) \in \mathfrak{R}$ also.

This definition could have been written as: (i) $\mathfrak{R}$ is said to be *reflexive* if, and only if, $x\mathfrak{R}x$ for all $x \in S$, (ii) $\mathfrak{R}$ is said to be *symmetric* if, and only if, for all $x, y \in S$ such that $x\mathfrak{R}y$, $y\mathfrak{R}x$ also, and (iii) $\mathfrak{R}$ is said to be *transitive* if, and only if, for all $x, y, z \in S$ such that $x\mathfrak{R}y$ and $y\mathfrak{R}z$, $x\mathfrak{R}z$ also.

Let us now examine the relation $\mathfrak{R}$ on the set J whose membership is displayed in Problem 3 of Exercise Group 6–1, to see if it possesses one or more of these properties. As this problem illustrates, a relation sometimes may be described verbally in several different ways; "is older than" is one way to describe this relation. We see that $\mathfrak{R}$ is not reflexive, since for no $x \in J$ is it true that x "is older than" x—in other words, for no $x \in J$ does $(x, x) \in \mathfrak{R}$. We also see that $\mathfrak{R}$ is not symmetric, since no $x, y \in J$ such that $(x, y) \in \mathfrak{R}$ is it true that $(y, x) \in \mathfrak{R}$; that is, if $x, y \in J$ and x is older than y it cannot possibly be true that y is older than x. We do note, however, that $\mathfrak{R}$ is transitive because, for any members $x, y, z \in J$ such that $(x, y), (y, z) \in \mathfrak{R}$, it is true that $(x, z) \in \mathfrak{R}$; that is, if $x, y, z \in J$, x is older than y, and y is older than z then, certainly, x is older than z. This relation is an example of an *order* relation. In Section 6–5 we shall learn what feature distinguishes order relations from other relations.

We now consider the relation on the set $S = \{0, 1, 2, 3\}$ mentioned in part (b) of Problem 1 of Exercise Group 6–1. $\mathfrak{R}$ was described as that relation on S such that, for $a, b \in S$, $a\mathfrak{R}b$ if and only if $a \cdot b = 2$. We see that

$$\mathfrak{R} = \{(1, 2), (2, 1)\}.$$

This relation is *not* reflexive, for there is no $x \in S$ such that $(x, x) \in \mathfrak{R}$; that is, there is no $x \in S$ such that $x \cdot x = 2$. This relation is, however, symmetric, because if $x, y \in S$ and $(x, y) \in \mathfrak{R}$, then $(y, x) \in \mathfrak{R}$ also; in other words, for any elements $x, y \in S$ such that $x \cdot y = 2$, then $y \cdot x = 2$. $\mathfrak{R}$ is *not* transitive, for we see that although $(1, 2) \in \mathfrak{R}$ and $(2, 1) \in \mathfrak{R}$, it is not true that $(1, 1) \in \mathfrak{R}$.

In each of the examples just given the relation possessed one of the properties in which we are interested. We now consider a relation which, at first glance, *appears* to possess two of the properties, the relation $\Re$ ("is a sibling of") on the set J (the Jones family), where

$$\Re = \begin{cases} (c,\,e),\ (c,\,p_1),\ (c,\,p_2),\ (c,\,t),\ (e,\,c),\ (e,\,p),\ (e,\,p_2), \\ (e,\,t),\ (p_1,\,c),\ (p_1,\,e),\ (p_1,\,p_2),\ (p_1,\,t),\ (p_2,\,c),\ (p_2,\,e), \\ (p_2,\,p_1),\ (p_2,\,t),\ (t,\,c),\ (t,\,e),\ (t,\,p_1),\ (t,\,p_2) \end{cases}.$$

$\Re$ is clearly not reflexive, for there is no $x \in J$ such that x is a sibling of x. However, $\Re$ is symmetric, for if $x,\,y \in J$ and $(x,\,y) \in \Re$, then $(y,\,x) \in \Re$. $\Re$ is not transitive, for if it were, $\Re$ would also be reflexive (why?).

EXERCISE GROUP 6–2

1. Consider the relation "is of different sex than" on the set J (the Jones family). Remembering that this relation is a subset of $J \times J$, list its complete membership. Determine which, if any, of Properties (i), (ii), (iii) are possessed by this relation.

2. Let B be the set of billionaires in your mathematics class. Determine whether the following statement is true or false: for all $x \in B$, x will receive an A in mathematics. Now consider the relation $\Re_1$ ("is a parent of") on the set J whose membership is displayed in Section 6–1. Show that $\Re_1$ is transitive.

6–3 Equivalence relations. We shall now discuss relations which possess *all three* of the properties (reflexivity, symmetry, and transitivity). Such relations are called *equivalence relations*. To show that there are such relations, let A be the set of all automobiles on the campus parking lot and let $\Re$ be a relation on A, where $\Re$ means "is the same make as." In this case, instead of writing out the membership of A or that of the relation $\Re$, we will assume that this has been done. After writing out the membership of A, we would display the membership of $A \times A$, and from this membership we would pick out those, and only those, ordered pairs $(x,\,y)$ for which the statement form $x\Re y$ (x is the same make as y) becomes a true statement. These ordered pairs, then, would make up the membership of the relation $\Re$ (subset of $A \times A$) and we can picture such a membership list in our mind's eye. Certainly $\Re$ is reflexive, since for each $x \in A$ it is true that $(x,\,x) \in \Re$ (x is the same make as x). $\Re$ is also symmetric, for if $x,\,y \in A$ and $(x,\,y) \in \Re$, then $(y,\,x) \in \Re$. $\Re$ is likewise transitive, for if $x,\,y,\,z \in A$ and $(x,\,y),\,(y,\,z) \in \Re$, then $(x,\,z) \in \Re$. Certainly, then, $\Re$ (is the same make as) is an equivalence relation on A. Problem 2 of Exercise Group 6–1 concerned writing out the membership of the relation $\Re$ (has the same birth month as) on the set S, where S was a convenient subset of the set of students in a mathematics class. This relation is also an equivalence relation: (i) each student in S has the same birth month as himself; (ii) if a first student in S has the same birth month as a second student in S, then the second student has the same birth month as the first student; and (iii) if a first student in S has the same birth month as a second student in S and the second student has the same birth month as a third student in S, then the first student has the same birth month as the third student.

EXERCISE GROUP 6–3

1. Let M be the set of all Minnesota citizens and let $\mathcal{R}$ mean "lives in the same county as." Show by a verbal argument that $\mathcal{R}$ is an equivalence relation.

2. Let D be the set of all residents of Duluth and let $\mathcal{R}$ mean "has the same surname as." Show by a verbal argument that $\mathcal{R}$ is an equivalence relation.

3. Let L be the set of all straight lines that may be drawn on a plane and let $\|$ mean "is parallel to." Is $\|$ an equivalence relation on L? Why? Let $\perp$ mean "is perpendicular to." Is $\perp$ an equivalence relation on L? Why?

4. Let F be the set of all items of furniture in a certain school building (desks, chairs, tables, globes, bookcases, etc.) and let $\mathcal{R}$ mean "belongs in the same room as." Show that $\mathcal{R}$ is an equivalence relation.

6–4 Partitions. Through the concept of "a partition of a set" we will come to an understanding of why we are interested in equivalence relations. We are all familiar with the fact that when a carpenter builds a house he first erects the exterior framework and then he "partitions" the house into rooms. But just what do we mean by saying that the house "is partitioned into rooms"? We know that the interior of the house contains a certain number of cubic feet of space, and we can visualize these "cubic feet" as being packed into the interior of the house in the same way that a child's blocks are packed away in a box. Strictly speaking, when we say that the house is *partitioned* we mean that the set of all cubic feet of space inside the house is divided into disjoint subsets. This homely example leads us to a formal definition of a partition of a set:

DEFINITION. Let S be any nonempty set and let $\mathcal{S} = \{T, U, V, \ldots\}$ be a family of nonempty subsets of S. The family $\mathcal{S}$ is called a *partition* of S if, and only if, (i) S is the union of $T, U, V, \ldots$ and (ii) any two distinct members of $\mathcal{S}$ are disjoint.

It is important to observe that nothing has been said to imply that $T, U, V, \ldots$ are distinct subsets of S; the definition does say, however, that if $\mathcal{S}$ is a partition of S then any two members of $\mathcal{S}$ are either identical (not distinct) or disjoint. This raises the possibility that some subset of S might appear in the membership list of $\mathcal{S}$ more than once, under different labels; for example, it might be true that the symbols T and U are labels for the same subset of S. Later we shall encounter partitions each element of which appears in the membership list infinitely many times but under different labels. For the present, however, we will content ourselves with more simple examples.

Consider the set A of all automobiles on the campus parking lot. If F is the set of all Fords on the parking lot and F' is the set of all non-Fords, a partition of A is $\mathfrak{A} = \{F, F'\}$, for F and F' are nonempty subsets of A, $F \cup F' = A$, and $F \cap F' = \emptyset$. If C is the set of all Chevrolets on the parking lot and C' is the set of all non-Chevrolets, another partition of A is the family $\mathfrak{B} = \{F, C, F' \cap C'\}$. (Explain!)

Suppose we are given a nonempty set S and a partition

$$\mathfrak{S} = \{T, U, V, \ldots\}$$

of S. Since S is the union of $T, U, V, \ldots$, any element $a \in S$ must belong to *at least one* of the subsets $T, U, V, \ldots$. But since any two subsets in the partition are either disjoint or identical, for each $a \in S$ there is a *unique* subset $X \in \mathfrak{S}$ such that $a \in X$. These observations lead us to conclude that the partition $\mathfrak{S}$ defines or "induces" a relation $\mathfrak{R}$ on the set S, where $\mathfrak{R}$ means "belongs to the same member of $\mathfrak{S}$ as" or "belongs to the same subset of the partition as." The important thing to note about this particular relation is that *it is an equivalence relation.*

To be more precise, the relation $\mathfrak{R}$ (belongs to the same subset of the partition as) is that relation on S such that for $a, b \in S$, $a\mathfrak{R}b$ if, and only if, there is a subset $X \in \mathfrak{S}$ such that $a, b \in X$. This relation is reflexive for $a\mathfrak{R}a$ for each $a \in S$. The relation $\mathfrak{R}$ is certainly symmetric, for if $a, b \in S$ and $a\mathfrak{R}b$, it is certainly true that $b\mathfrak{R}a$. To show that $\mathfrak{R}$ is transitive, suppose that $a, b, c \in S$, $a\mathfrak{R}b$, and $b\mathfrak{R}c$. Then, since $a\mathfrak{R}b$, there is a subset X of the partition (an element of $\mathfrak{S}$) such that $a, b \in X$, and since $b\mathfrak{R}c$, there is a subset Y of the partition such that $b, c \in Y$. Thus, since $b \in X$ and $b \in Y$, $X \cap Y$ is not empty, X and Y are *not distinct* subsets of the partition (if they were distinct they would be disjoint), and $X = Y$. Thus $a \in Y$, and therefore $a\mathfrak{R}c$ and $\mathfrak{R}$ is proved to be transitive. We conclude, then, that $\mathfrak{R}$ (belongs to the same subset of the partition as) is an equivalence relation.

Whenever we are given any nonempty set S together with a partition $\mathfrak{S}$ of S, we know that the relation $\mathfrak{R}$ (belongs to the same element of $\mathfrak{S}$ as) is an equivalence relation on S. In other words, if we are given a partition of any set S we need never be at loss to find an equivalence relation on S. As an example, let S be the set of students in your mathematics class and let

$$\mathfrak{S} = \{A, B, C, D, E, F, G, H, I, J\}$$

be a partition of S, where the subset A of S consists of members whose age, to the nearest year, ends with the digit 0; the subset B of S consists of members whose age, to the nearest year, ends with the digit 1; and so on. Then an obvious equivalence relation $\Re$ on S is "has age to the nearest year ending with the same digit as" or "belongs to the same element of $\mathcal{S}$ as." (Show that this is "obvious.") In this example we are tacitly assuming that the sets $A, B, C, \ldots$ are nonempty. If, for example, it should happen that there are no students in your mathematics class whose age to the nearest year ends with the digit 0, the symbol A should not be entered in the membership list of $\mathcal{S}$.

Not only does a given partition of a nonempty set S induce a natural equivalence relation on S but, conversely, a given equivalence relation on S induces (defines) a natural way of partitioning S into subsets which we call *equivalence sets*. To prove this assertion, let us consider a nonempty set

$$S = \{a, b, c, d, \ldots\},$$

and let $\Re$ be some given equivalence relation on S. For each element $u \in S$ let S_u (S sub u) be the subset of S containing those, and only those, elements $x \in S$ such that $x\Re u$. We shall now show that the family of subsets

$$\mathcal{S} = \{S_a, S_b, S_c, S_d, \ldots\}$$

is a partition of S, where S_a is the set of those elements $x \in S$ such that $x\Re a$, S_b is the set of those elements $x \in S$ such that $x\Re b$, and so on. First we prove that

$$S = S_a \cup S_b \cup S_c \cup S_d \cup \ldots$$

To do this, we take any element $u \in S$. Since $\Re$ is reflexive and $u\Re u$ it follows that $u \in S_u$, so that u belongs to at least one of the sets of the union on the right. Therefore S must be a subset of the set on the right. On the other hand, if we take an element v belonging to the union on the right, v will belong to *at least one* of the sets $S_a, S_b, S_c, S_d, \ldots$. Since each of these sets contains only elements which belong to S, it follows that $v \in S$, and we have shown that the union on the right is a subset of S on the left. Therefore, since the sets on the right and left include each other, they are equal; the family $\mathcal{S} = \{S_a, S_b, S_c, S_d, \ldots\}$ satisfies one of the requirements for being a partition of S.

We next show that any two subsets of S belonging to the family $\mathcal{S}$ are either disjoint or identical. To do this, suppose that S_u and S_v are any two elements of $\mathcal{S}$ and that S_u and S_v are *not disjoint*; that is, $S_u \cap S_v$ contains at least one element x. Since $x \in S_u$, $x\Re u$, and since $x \in S_v$, $x\Re v$. Now, because $\Re$ is symmetric, $u\Re x$, and, since $\Re$ is transitive, $u\Re v$.

We now take any element $y \in S_u$ so that $y\mathcal{R}u$ and, since $u\mathcal{R}v$ and $\mathcal{R}$ is transitive, it follows that $y\mathcal{R}v$. Hence, $y \in S_v$ and $S_u \subset S_v$.

On the other hand, taking any element $z \in S_v$, we have $z\mathcal{R}v$. Thus, since $v\mathcal{R}u$ ($\mathcal{R}$ is symmetric) and $\mathcal{R}$ is transitive, it follows that $z\mathcal{R}u$, so that $z \in S_u$ and $S_v \subset S_u$. Therefore, because S_u and S_v include each other, $S_u = S_v$ and we have shown that any two elements of $\mathcal{S}$ which are not disjoint are identical, i.e., the family $\mathcal{S} = \{S_a, S_b, S_c, S_d, \ldots\}$ satisfies the second requirement for being a partition of S. We shall call the partition $\mathcal{S}$ "the partition of S induced by the equivalence relation $\mathcal{R}$."

It is important to note that we may scan the membership of $\mathcal{S}$ and find the same subset appearing more than once. As an example, let S be the set of students in your mathematics class and let $\mathcal{R}$ be the equivalence relation "sits in the same row as." If u and v are students in the set S who sit in the same row, then S_u and S_v will be exactly the same subsets.

We have shown that any partition $\mathcal{S}$ of a nonempty set S induces an equivalence relation $\mathcal{R}$ on S, and we have also shown that any equivalence relation $\mathcal{R}$ on S induces a partition $\mathcal{S}$ of S. As a matter of interest, we merely mention that there is a $1:1$ correspondence between the set of all equivalence relations on S and the set of all partitions of S; you are invited to prove this fact in the second problem of the exercises which follow.

EXERCISE GROUP 6–4

1. Consider the set $S = \{a, b, c, d, e, f, g\}$, where a, b, c, d, e, f, and g are students in a classroom. Assume that the students are seated in rows according to the following scheme.

3rd row:	f	g	
2nd row:	c	d	e
1st row:	a	b	

Write out the membership list for the sets S_a, S_b, S_c, S_d, S_e, S_f, S_g of the partition of S induced by the equivalence relation $\mathcal{R}$ (sits in the same row as).

If a, e, f are women and b, c, d, g are men, do the sets

$$S_a = \{a, e, f\}, \quad S_b = \{b, c, d, g\}, \quad S_c = \{b, c, d, g\}, \quad S_d = \{b, c, d, g\},$$
$$S_e = \{a, e, f\}, \quad S_f = \{a, e, f\}, \quad S_g = \{b, c, d, g\}$$

form a partition of S? Explain. If these sets do form a partition of S, what equivalence relation on S does this partition define?

2. Prove that the set of all equivalence relations on a nonempty set S may be put into $1:1$ correspondence with the set of all partitions of S. The following suggestions are offered:

    (a) Let R denote the set of equivalence relations and let P denote the set of partitions.

(b) Let α be the mapping of R into P which associates each $\mathcal{R} \in R$ with its induced partition of S.

(c) Show that α is a mapping of R *onto* P; that is, show that for each $\mathcal{P} \in P$ there is a relation $\mathcal{R} \in R$ such that $\alpha : \mathcal{R} \to \mathcal{P}$.

(d) Show that α maps distinct relations into distinct partitions.

6–5 Order relations. An example of an order relation was introduced in Section 6–2, but since order relations will be important in our later work, we shall study them in more detail. We start with two definitions:

DEFINITION. (i) A relation $\mathcal{R}$ on a set S is said to be *antireflexive* if, and only if, $(x, x) \notin \mathcal{R}$ for *all* $x \in S$, and (ii) a relation $\mathcal{R}$ on a set S is said to be *antisymmetric* if, and only if, whenever $(x, y) \in \mathcal{R}$, then $(y, x) \notin \mathcal{R}$.

With these new adjectives in our vocabulary, we are ready to define an order relation:

DEFINITION. If $\mathcal{R}$ is a relation on a set S, we call $\mathcal{R}$ an *order relation* if, and only if, $\mathcal{R}$ is antireflexive, antisymmetric, and transitive.

An immediate example of an order relation is the relation $<$ (is less than) on the set of natural numbers. Recalling our definition of a relation, we know that $<$ is really a subset of $N \times N$, and we can say that $<$ contains those, and only those, ordered pairs of natural numbers each of which has the property that its left partner "is less than" its right partner. Property (ix) of the natural numbers (proved in Section 2–2) assures us that the relation $<$ is transitive. However, for any $a \in N$ it is not true that $a < a$ (why?) and for any $a, b \in N$ such that $a < b$ it is not true that $b < a$ (why?). Thus we see that $<$ is antireflexive, antisymmetric, and transitive, and therefore is an order relation on the natural numbers.

As a further example of an order relation on N, consider the set $T = \{3, 6, 9, 12, 15, \ldots\}$ of all nonzero natural numbers which are multiples of 3 $(1 \cdot 3, 2 \cdot 3, 3 \cdot 3, 4 \cdot 3, \ldots)$ and let $\mathcal{R}$ be the subset of $N \times N$ (relation on N) which contains those, and only those, ordered pairs of natural numbers whose left partners exceed their corresponding right partners by a nonzero multiple of 3. In other words, $\mathcal{R}$ is the subset of $N \times N$ such that, for any $(a, b) \in N \times N$, $(a, b) \in \mathcal{R}$ if, and only if, $a = b + t$ for some $t \in T$. For example, the pairs $(6, 0)$, $(7, 1)$, $(11, 2)$ are members of $\mathcal{R}$ but $(2, 5)$, $(9, 4)$, $(23, 15)$ are not.

The relation $\mathcal{R}$ is clearly antireflexive, since $(a, a) \notin \mathcal{R}$ for all $a \in N$. $\mathcal{R}$ is also antisymmetric, for if $(a, b) \in \mathcal{R}$ so that $a = b + t$ for some $t \in T$, it follows that $b < a$ and $(b, a) \notin \mathcal{R}$. To prove that $\mathcal{R}$ is transitive we make use of the fact that the sum of any two elements of T is also an element of T: if $m \cdot 3, n \cdot 3 \in T$, then $m \cdot 3 + n \cdot 3 = (m + n) \cdot 3 \in T$.

Thus, if (a, b), $(b, c) \in \mathfrak{R}$, there is an element $t_1 \in T$ such that $a = b + t_1$, and there is also an element $t_2 \in T$ such that $b = c + t_2$. Then, by using the substitution principle together with Axiom III, we obtain

$$a = (c + t_2) + t_1 = c + (t_2 + t_1),$$

which proves that $(a, c) \in \mathfrak{R}$ and $\mathfrak{R}$ is transitive. This example should suggest a method for finding many other order relations on the natural numbers.

CHAPTER 7

THE INTEGERS

7-1 The relation $\sim$ on $N \times N$. The system $\{N; +, \cdot\}$ of natural numbers has a number of weaknesses, perhaps the most glaring of which is that subtraction is not an operation on the set N of natural numbers, i.e., given any two natural numbers it is not always possible to subtract one from the other. Briefly, we may say that $\{N; +\}$ is not a group, for each natural number does not have an additive inverse. We intend to remedy this situation, not by somehow endowing the natural numbers with additional properties, but by constructing a new number system called the *integers*. It will turn out that this new number system will (in a certain sense) have the set N of natural numbers as a subset, and the natural numbers themselves will later be called the *non-negative integers*. When we have finished constructing the integers we will say that we have *extended* the natural numbers and that the natural numbers are *embedded* in the integers. The process by which we extend the natural numbers is part of the process of orderly and logical growth mentioned in Section 1–3 (which should be reviewed at this time).

From our definition of a number system (review Section 4–5), we realize that we must find a set and define two operations on it in such a way that both operations are commutative and associative and one is distributive with respect to the other. How are we to find a set to start with? Since we want the number system that we are going to construct to be an extension of the natural number system, we will construct our set by using the natural numbers as building blocks, as follows. First of all, let us consider the set $N \times N$ of all ordered pairs of natural numbers. This is not a difficult set to visualize but the membership list is awkward to write. We write it below as a rectangular array of ordered pairs:

$$N \times N = \begin{Bmatrix} (0, 0), & (0, 1), & (0, 2), & (0, 3), \ldots \\ (1, 0), & (1, 1), & (1, 2), & (1, 3), \ldots \\ (2, 0), & (2, 1), & (2, 2), & (2, 3), \ldots \\ (3, 0), & (3, 1), & (3, 2), & (3, 3), \ldots \\ \vdots & \vdots & \vdots & \vdots \end{Bmatrix}.$$

We now define a relation $\sim$ (curl) on the set $N \times N$ as follows:

DEFINITION. If (a, b), $(c, d) \in N \times N$, then $(a, b) \sim (c, d)$ if, and only if, $a + d = b + c$. We read $(a, b) \sim (c, d)$ "the pair (a, b) has the curl relation to the pair (c, d)."

Strictly speaking, the relation $\sim$ on $N \times N$ is a subset of $(N \times N) \times (N \times N)$, where each member of the relation $\sim$ is an ordered pair, each of whose partners is an ordered pair of natural numbers. We will avoid trying to display the membership of the relation $\sim$. We note, however, that $[(3, 2), (5, 4)]$ is a member of $\sim$ because $3 + 4 = 2 + 5$. The ordered pairs $[(2, 7), (6, 11)]$ and $[(4, 1), (8, 5)]$ are also members of the relation $\sim$. (We have used brackets to enclose the members of $\sim$, since parentheses have been used to enclose the partners of each member.) We now show that the relation $\sim$ on $N \times N$ is an equivalence relation:

(i) $\sim$ is *reflexive*, for if (a, b) is any member of $N \times N$, then $(a, b) \sim (a, b)$, since $a + b = b + a$.

(ii) $\sim$ is *symmetric*, for if (a, b), (c, d) are any members of $N \times N$ such that $(a, b) \sim (c, d)$, then $a + d = b + c$, from which it follows that $c + b = d + a$ and $(c, d) \sim (a, b)$.

(iii) To show that $\sim$ is *transitive* let (a, b), (c, d), (e, f) be any members of $N \times N$ such that $(a, b) \sim (c, d)$ and $(c, d) \sim (e, f)$. Then, since $(a, b) \sim (c, d)$ and $(c, d) \sim (e, f)$, we know that $a + d = b + c$ and $c + f = d + e$. Adding f to both sides [Property (iii)] of the first equation and using the associative property gives us

$$(a + d) + f = (b + c) + f = b + (c + f),$$

and adding b to both sides of the second equation gives us

$$b + (c + f) = b + (d + e).$$

Therefore, by the substitution principle, we have

$$(a + d) + f = b + (d + e).$$

Finally, by using the associative, commutative, and cancellation axioms as they apply to addition of natural numbers, we can write

$$(a + d) + f = b + (d + e),$$
$$a + (d + f) = b + (e + d),$$
$$a + (f + d) = b + (e + d),$$
$$(a + f) + d = (b + e) + d,$$
$$a + f = b + e,$$

so that $(a, b) \sim (e, f)$, as we wished to show.

Now that we know the relation $\sim$ to be an equivalence relation, we can use it to partition the set $N \times N$ (review Section 6-4). For each pair $(a, b) \in N \times N$ let $I(a, b)$ (read "I of a and b") represent the set of all ordered pairs $(x, y) \in N \times N$ such that $(x, y) \curvearrowleft (a, b)$. The family I of all sets of the form $I(a, b)$, $(a, b) \in N \times N$, is the partition of $N \times N$ induced by the equivalence relation $\curvearrowleft$. We list the membership of I in array form:

$$I = \begin{cases} I(0, 0), & I(0, 1), & I(0, 2), & I(0, 3), \ldots \\ I(1, 0), & I(1, 1), & I(1, 2), & I(1, 3), \ldots \\ I(2, 0), & I(2, 1), & I(2, 2), & I(2, 3), \ldots \\ I(3, 0), & I(3, 1), & I(3, 2), & I(3, 3), \ldots \\ \vdots & \vdots & \vdots & \vdots \end{cases}$$

We know that any two members of I are either disjoint or identical and that the union of all the members of I is $N \times N$; that is, I is a partition of $N \times N$. We need not prove this, for it has already been proved by our general discussion in Section 6–4.

7–2 The operations $\oplus$ and $\otimes$ on I. The members of the set I will be called *integers*, and since we want these integers to eventually be the numbers in a new number system, our first task is to define operations on the set I. We shall define two operations on I: *addition*, for which we use the symbol $\oplus$, and *multiplication* for which we use the symbol $\otimes$. We are aware that the symbols in the membership list of I represent subsets of $N \times N$ and we are also aware that two different symbols in this membership list might possibly represent the same subset. Therefore we must define addition and multiplication in such a way that the sum or product obtained by adding or multiplying any two integers is independent of the symbols we use for these integers; in short, addition and multiplication must be *well defined*. We state the following theorem:

THEOREM. If $I(a, b) = I(e, f)$ and $I(c, d) = I(g, h)$, then

$$I(a + c, b + d) = I(e + g, f + h), \tag{i}$$

and

$$I(a \cdot c + b \cdot d, a \cdot d + b \cdot c) = I(e \cdot g + f \cdot h, e \cdot h + f \cdot g). \tag{ii}$$

This theorem guarantees that addition and multiplication (as we are about to define them) are well defined. The proof will be given in Section 7–5 after a more manageable set of symbols for the integers has been introduced, but you are invited to prove it for yourself at this time. We now define addition:

DEFINITION. If $I(a, b)$ and $I(c, d)$ are any integers (members of I), then $I(a, b) \oplus I(c, d) = I(a + c, b + d)$.

Remember that this is a *definition* for addition of integers. This definition associates each ordered pair of integers with another integer, i.e., it is a rule for mapping $I \times I$ into I. For example, $I(3, 2) \oplus I(2, 5) = I(5, 7)$ and $I(0, 4) \oplus I(8, 3) = I(8, 7)$.

Our next operation is multiplication, for which we use the symbol $\otimes$ (times) and which we define as follows:

DEFINITION. If $I(a, b)$ and $I(c, d)$ are any integers, then

$$I(a, b) \otimes I(c, d) = I(a \cdot c + b \cdot d, \, a \cdot d + b \cdot c).$$

For example,

$$I(2, 1) \otimes I(3, 2) = I(2 \cdot 3 + 1 \cdot 2, \, 2 \cdot 2 + 1 \cdot 3) = I(8, 7),$$

and

$$I(0, 2) \otimes I(4, 0) = I(0 \cdot 4 + 2 \cdot 0, \, 0 \cdot 0 + 2 \cdot 4) = I(0, 8).$$

The reader might complain, with some justification, that these definitions of "addition" and "multiplication" seem artificial and unfamiliar. However, the rules for adding and multiplying integers learned in elementary high-school algebra are actually artificial rules, although they no longer seem so because we have learned them so well. Here we shall attempt to prove that they are valid.

7–3 The commutativity and associativity of $\oplus$ and $\otimes$. The operation $\oplus$ is both commutative and associative on the set I of integers, and this is very easy to prove. If $I(a, b)$ and $I(c, d)$ are any integers, then

$$I(a, b) \oplus I(c, d) = I(a + c, \, b + d),$$

and

$$I(c, d) \oplus I(a, b) = I(c + a, \, d + b).$$

Since $a + c = c + a$ and $b + d = d + b$,

$$I(a + c, \, b + d) = I(c + a, \, d + b),$$

and addition is proved *commutative*. If $I(a, b)$, $I(c, d)$, and $I(e, f)$ are any integers, then

$$I(a, b) \oplus [I(c, d) \oplus I(e, f)]$$
$$= I(a, b) \oplus I(c + e, \, d + f) = I(a + [c + e], \, b + [d + f]),$$

and

$$[I(a, b) \oplus I(c, d)] \oplus I(e, f)$$
$$= I(a + c, \, b + d) \oplus I(e, f) = I([a + c] + e, \, [b + d] + f).$$

Since

$$a + [c + e] = [a + c] + e \quad \text{and} \quad b + [d + f] = [b + d] + f,$$
$$I(a + [c + e], \ b + [d + f]) = I([a + c] + e, \ [b + d] + f),$$

and addition is proved *associative*. It is important to appreciate that the natural numbers, in a roundabout way, were used to construct the integers; the natural numbers may be thought of as *ancestors* and the integers as *offspring*. The commutativity and associativity of addition on the integers may be thought of as being *inherited*.

To show that multiplication of integers is commutative we observe that if $I(a, b)$ and $I(c, d)$ are any integers,

$$I(a, b) \otimes I(c, d) = I(a \cdot c + b \cdot d, \ a \cdot d + b \cdot c);$$

and

$$I(c, d) \otimes I(a, b) = I(c \cdot a + d \cdot b, \ c \cdot b + d \cdot a).$$

Since

$$a \cdot c + b \cdot d = c \cdot a + d \cdot b \quad \text{and} \quad a \cdot d + b \cdot c = c \cdot b + d \cdot a,$$

it follows that

$$I(a \cdot c + b \cdot d, \ a \cdot d + b \cdot c) = I(c \cdot a + d \cdot b, \ c \cdot b + d \cdot a),$$

and multiplication is proved commutative. To prove that multiplication is associative we consider any integers $I(a, b)$, $I(c, d)$, and $I(e, f)$ and compare the products

$$I(a, b) \otimes [I(c, d) \otimes I(e, f)] \quad \text{and} \quad [I(a, b) \otimes I(c, d)] \otimes I(e, f).$$

The first product is

$$I(a, b) \otimes I(c \cdot e + d \cdot f, \ c \cdot f + d \cdot e)$$
$$= I(a \cdot [c \cdot e + d \cdot f] + b \cdot [c \cdot f + d \cdot e], \ a \cdot [c \cdot f + d \cdot e] + b \cdot [c \cdot e + d \cdot f]),$$

and the second product is

$$I(a \cdot c + b \cdot d, \ a \cdot d + b \cdot c) \otimes I(e, f)$$
$$= I([a \cdot c + b \cdot d] \cdot e + [a \cdot d + b \cdot c] \cdot f, \ [a \cdot c + b \cdot d] \cdot f + [a \cdot d + b \cdot c] \cdot e).$$

Because multiplication of natural numbers is associative and distributive over addition and because addition of natural numbers is commutative and associative, we see that

$$a \cdot [c \cdot e + d \cdot f] + b \cdot [c \cdot f + d \cdot e] = [a \cdot c + b \cdot d] \cdot e + [a \cdot d + b \cdot c] \cdot f,$$

and

$$a \cdot [c \cdot f + d \cdot e] + b \cdot [c \cdot e + d \cdot f] = [a \cdot c + b \cdot d] \cdot f + [a \cdot d + b \cdot c] \cdot e,$$

so that the two products are equal; multiplication of integers is therefore associative, as we wished to show.

7–4 The number system $\{I; \oplus, \otimes\}$. In the previous two sections we have defined two operations on the set I of integers and have shown that each is both commutative and associative. If we can show that one of these operations is distributive with respect to the other we will be able to conclude that the system $\{I; \oplus, \otimes\}$ is a number system. We now propose to show that multiplication on the integers is distributive with respect to addition; that is, if $I(a, b)$, $I(c, d)$, $I(e, f)$ are any integers, then

$$I(a, b) \otimes [I(c, d) \oplus I(e, f)] = I(a, b) \otimes I(c, d) \oplus I(a, b) \otimes I(e, f).$$

If we work out the left side we do addition first, followed by multiplication, and obtain

$$
\begin{aligned}
I(a, b) &\otimes [I(c, d) \oplus I(e, f)] \\
&= I(a, b) \otimes I(c + e, d + f) \\
&= I(a \cdot [c + e] + b \cdot [d + f], \, a \cdot [d + f] + b \cdot [c + e]).
\end{aligned}
$$

In working out the right side we do the two multiplications first, followed by the addition of the products, and obtain

$$
\begin{aligned}
I(a, b) &\otimes I(c, d) \oplus I(a, b) \otimes I(e, f) \\
&= I(a \cdot c + b \cdot d, \, a \cdot d + b \cdot c) \oplus I(a \cdot e + b \cdot f, \, a \cdot f + b \cdot e) \\
&= I([a \cdot c + b \cdot d] + [a \cdot e + b \cdot f], \, [a \cdot d + b \cdot c] + [a \cdot f + b \cdot e]).
\end{aligned}
$$

We digress for a moment to recall that an integer $I(u, v)$ is a set of all ordered pairs $(x, y) \in N \times N$ such that $(x, y) \curvearrowright (u, v)$. It is commonly said that the pair (u, v) "identifies" the integer (equivalence set) $I(u, v)$. Now, in order to prove the distributive property of multiplication, we must show that

$$
\begin{aligned}
I(a \cdot [c + e] &+ b \cdot [d + f], \, a \cdot [d + f] + b \cdot [c + e]) \\
&= I([a \cdot c + b \cdot d] + [a \cdot e + b \cdot f], \, [a \cdot d + b \cdot c] + [a \cdot f + b \cdot e]),
\end{aligned}
$$

and to show that these integers are equal we show that they are both "identified" by the same pair of natural numbers. Observe that

$$a \cdot [c + e] + b \cdot [d + f] = [a \cdot c + a \cdot e] + [b \cdot d + b \cdot f] \qquad \text{(Why?)}$$
$$= ([a \cdot c + a \cdot e] + b \cdot d) + b \cdot f \qquad \text{(Why?)}$$
$$= (a \cdot c + [a \cdot e + b \cdot d]) + b \cdot f \qquad \text{(Why?)}$$
$$= (a \cdot c + [b \cdot d + a \cdot e]) + b \cdot f \qquad \text{(Why?)}$$
$$= ([a \cdot c + b \cdot d] + a \cdot e) + b \cdot f \qquad \text{(Why?)}$$
$$= [a \cdot c + b \cdot d] + [a \cdot e + b \cdot f], \qquad \text{(Why?)}$$

so that the left partners of the identifying pairs are the same. The right partners of these pairs may similarly be proved equal (do it!) so that, since the identifying pairs are the same, the integers are equal and *multiplication of integers is distributive over addition*. We have now attained our immediate goal so far as the integers are concerned: *we have shown that the system* $\{I; \oplus, \otimes\}$ *is a number system*.

EXERCISE GROUP 7–1

1. If a, b, c, d, e, f are natural numbers, prove that

$$a \cdot [c \cdot e + d \cdot f] + b \cdot [c \cdot f + d \cdot e] = [a \cdot c + b \cdot d] \cdot e + [a \cdot d + b \cdot c] \cdot f.$$

You will need to use the associative and distributive properties of multiplication, together with the commutative and associative properties of addition. Work with *one side only*.

2. Perform the following operations:

    (a) $[I(3, 1) \oplus I(5, 8)] \oplus I(8, 3) = \underline{\hspace{1.5cm}}$

    (b) $I(4, 3) \otimes [I(1, 2) \oplus I(4, 2)] = \underline{\hspace{1.5cm}}$

    (c) $[I(7, 4) \otimes I(5, 3)] \otimes I(2, 2) = \underline{\hspace{1.5cm}}$

3. Prove that the integers (sets) $I(2, 5)$ and $I(4, 7)$ are equal by actually proving that $I(2, 5) \subset I(4, 7)$ and $I(4, 7) \subset I(2, 5)$.

4. If a, b, c, d, e, f are natural numbers, prove that

$$a \cdot [d + f] + b \cdot [c + e] = [a \cdot d + b \cdot c] + [a \cdot f + b \cdot e].$$

Work with *one side only*.

7–5 A new notation for the integers. In this section we shall attempt to find a new and more convenient notation for the elements of the set I whose members are displayed in array form in Section 7–1. We first ask if each element of I might not perhaps appear in the membership list more than once. If it turns out that such is the case, we can certainly simplify the listing of I's membership by entering each distinct integer only once. To see if this can be done let us look again at the set

$$N \times N = \left\{ \begin{array}{llll} (0,0), & (0,1), & (0,2), & (0,3), \ldots \\ (1,0), & (1,1), & (1,2), & (1,3), \ldots \\ (2,0), & (2,1), & (2,2), & (2,3), \ldots \\ (3,0), & (3,1), & (3,2), & (3,3), \ldots \\ \vdots & \vdots & \vdots & \vdots \end{array} \right\},$$

and let us look especially closely at the subsets of $N \times N$ whose elements lie along the diagonal lines. To facilitate the study of these subsets we adopt the following symbols for them:

$$
\begin{array}{lll}
\qquad \vdots \quad \vdots \quad \vdots \quad \vdots \qquad\quad \vdots & & \\
\{(0,3), \ (1,4), \ (2,5), \ (3,6), \ldots\} = -3 & \text{(minus 3)} \\
\{(0,2), \ (1,3), \ (2,4), \ (3,5), \ldots\} = -2 & \text{(minus 2)} \\
\{(0,1), \ (1,2), \ (2,3), \ (3,4), \ldots\} = -1 & \text{(minus 1)} \\
\{(0,0), \ (1,1), \ (2,2), \ (3,3), \ldots\} = \ \ 0 & \text{(zero)} \\
\{(1,0), \ (2,1), \ (3,2), \ (4,3), \ldots\} = +1 & \text{(plus 1)} \\
\{(2,0), \ (3,1), \ (4,2), \ (5,3), \ldots\} = +2 & \text{(plus 2)} \\
\{(3,0), \ (4,1), \ (5,2), \ (6,3), \ldots\} = +3 & \text{(plus 3)} \\
\qquad \vdots \quad \vdots \quad \vdots \quad \vdots \qquad\quad \vdots & &
\end{array}
$$

The symbols $\ldots -3, -2, -1, 0, +1, +2, +3, \ldots$ are merely symbols we have adopted to represent these diagonal sets and no other meaning is to be attached to them at this time.

Since each pair $(a, b) \in N \times N$ lies on one and only one diagonal, it is clear that the sets

$$\ldots -3, \ -2, \ -1, \ 0, \ +1, \ +2, \ +3, \ldots$$

are disjoint. Furthermore, the union of all the sets

$$\ldots -3, \ -2, \ -1, \ 0, \ +1, \ +2, \ +3, \ldots$$

is clearly $N \times N$, so that the family of sets

$$\{\ldots -3, \ -2, \ -1, \ 0, \ +1, \ +2, \ +3, \ldots\}$$

is a *partition* of $N \times N$ and as such *must define some equivalence relation on* $N \times N$. We now seek to discover what equivalence relation is induced on $N \times N$ by the partition

$$\{\ldots -3, \ -2, \ -1, \ 0, \ +1, \ +2, \ +3, \ldots\}$$

and to help us do so, we note that any diagonal set is of the form

$$\{(a, b), \ (a + 1, \ b + 1), \ (a + 2, \ b + 2), \ldots\},$$

where the leading pair (a, b) is the pair of natural numbers found on the top row or left column of the $N \times N$ membership array [at least one of the partners of (a, b) is 0]. Let us look at any two elements $(a + h, \ b + h)$ and $(a + k, \ b + k)$ from this diagonal set, where h and k are any natural numbers. If we add the left partner of the first pair to the right partner of the second, we obtain

$$(a + h) + (b + k) = [(a + h) + b] + k = [a + (h + b)] + k$$
$$= [a + (b + h)] + k = [(a + b) + h] + k = (a + b) + (h + k),$$

while if we add the right partner of the first pair to the left partner of the second, we obtain

$$(b + h) + (a + k) = (a + k) + (b + h) = [(a + k) + b] + h$$
$$= [a + (k + b)] + h = [a + (b + k)] + h = [(a + b) + k] + h$$
$$= (a + b) + (k + h) = (a + b) + (h + k).$$

Therefore it is clear that $(a + h, \ b + h) \cap (a + k, \ b + k)$, that is, each pair of natural numbers in the diagonal set has the relation $\cap$ to each other pair in the set. This is true, in particular, when $h = 0$, so that every pair in the set has the relation $\cap$ to the leading pair (a, b).

Let us now investigate any two disjoint diagonal sets of the $N \times N$ membership array. We will consider five cases:

(i) $+h = \{(h, 0), (h + 1, 1), \ldots\}$ and $+k = \{(k, 0), (k + 1, 1), \ldots\}$,

(ii) $+h = \{(h, 0), (h + 1, 1), \ldots\}$ and $0 = \{(0, 0), (1, 1), \ldots\}$,

(iii) $+h = \{(h, 0), (h + 1, 1), \ldots\}$ and $-k = \{(0, k), (1, k + 1), \ldots\}$,

(iv) $0 = \{(0, 0), (1, 1), \ldots\}$ and $-k = \{(0, k), (1, k + 1), \ldots\}$,

(v) $-h = \{(0, h), (1, h + 1), \ldots\}$ and $-k = \{(0, k), (1, k + 1), \ldots\}$.

We assume in all cases that $h \neq 0$, $k \neq 0$, and that $h \neq k$ except, possibly, in case (iii). In case (i), we see that $h + 0 \neq 0 + k$, so that $(h, 0) \not\cap (k, 0)$, where $\not\cap$ means "does *not* have the relation $\cap$ to"; in case (ii), $h + 0 \neq 0 + 0$, so that $(h, 0) \not\cap (0, 0)$; in case (iii), $h + k \neq$

$0 + 0$, so that $(h, 0) \not\sim (0, k)$; in case (iv), $0 + k \neq 0 + 0$, so that $(0, 0) \not\sim (0, k)$; and in case (v), $0 + k \neq h + 0$, so that $(0, h) \not\sim (0, k)$. We see by observing these five cases that the leading pairs of any two distinct diagonal sets do *not* have the relation $\sim$. Thus, since $\sim$ is a transitive relation, no pair belonging to one diagonal set ever has the relation $\sim$ to a pair belonging to another diagonal set. If we again consider any diagonal set

$$\{(a, b), (a + 1, b + 1), (a + 2, b + 2), \ldots\},$$

the observations that we have made above tell us that this set is the set of all $(x, y) \in N \times N$ such that $(x, y) \sim (a, b)$; it is the set of all $(x, y) \in N \times N$ such that $(x, y) \sim (a + 1, b + 1)$; it is the set of all $(x, y) \in N \times N$ such that $(x, y) \sim (a + 2, b + 2)$, and so on. In other words, this set is the set $I(a, b)$, $I(a + 1, b + 1)$, $I(a + 2, b + 2)$, and so on.

We look again at the set I with its membership listed in array form, as in Section 7–1:

$$I = \left\{ \begin{array}{l} I(0,0), \ I(0,1), \ I(0,2), \ I(0,3), \ldots \\ I(1,0), \ I(1,1), \ I(1,2), \ I(1,3), \ldots \\ I(2,0), \ I(2,1), \ I(2,2), \ I(2,3), \ldots \\ I(3,0), \ I(3,1), \ I(3,2), \ I(3,3), \ldots \\ \vdots \quad \vdots \quad \vdots \quad \vdots \end{array} \right\}.$$

If we look down any diagonal of integers we see that the identifying pairs of these integers are precisely the pairs of natural numbers lying along the corresponding diagonal of the $N \times N$ membership array. This means, of course, that all the integers lying on any one of the above diagonals are *the same*. For example, $I(0, 1)$, $I(1, 2)$, $I(2, 3)$, $\ldots$ are identical; the symbols $I(0, 1)$, $I(1, 2)$, $I(2, 3)$, $\ldots$ are merely different representations for the same integer. In fact, every integer has infinitely many representations, but we shall find it convenient to settle on *just one* representation for each integer and so we adopt the following convention: *Of the infinitely many representations for any integer, we shall call that representation whose identifying pair has at least one partner equal to 0 the standard representation.* Thus, using the standard representation for each integer, the membership list of I may be much more simply written as

$$I = \{\ldots, I(0, 3), \ I(0, 2), \ I(0, 1), \ I(0, 0), \ I(1, 0), \ I(2, 0), \ I(3, 0), \ldots\}.$$

In short, for each integer we have retained only one of its infinitely many representations. We must now be sure that we have not gone too far in this matter. If two integers are to be added or multiplied, will the sum or product be independent of the representation we use for the integers? If the answer to this question is "yes" we will know that when adding or multiplying integers we may be assured of obtaining a unique sum or product if standard representation for the integers is used.

To show that such is the case, suppose that $I(a, b)$ and $I(c, d)$ are standard representations for two integers. Then,

$$I(a, b) \oplus I(c, d) = I(a + c, b + d).$$

If, on the other hand, h and k are any natural numbers and the representations

$$I(a + h, b + h),$$
$$I(c + k, d + k)$$

are used, we obtain

$$I(a + h, b + h) \oplus I(c + k, d + k) = I([a + h] + [c + k],$$
$$[b + h] + [d + k])$$
$$= I([a + c] + [h + k],$$
$$[b + d] + [h + k]).$$

Since

$$(a + c, b + d) \cap ([a + c] + [h + k], [b + d] + [h + k]),$$

it is clear that

$$I(a + c, b + d) = I([a + c] + [h + k], [b + d] + [h + k]),$$

and the sum of any two integers is independent of the representation used. To show that this also holds true for multiplication, we find the products

$$I(a, b) \otimes I(c, d)$$

and

$$I(a + h, b + h) \otimes I(c + k, d + k).$$

The first product is $I(a \cdot c + b \cdot d, a \cdot d + b \cdot c)$ and the second is

$$I([a + h] \cdot [c + k] + [b + h] \cdot [d + k],$$
$$[a + h] \cdot [d + k] + [b + h] \cdot [c + k]).$$

Now let us observe that the left partner of the identifying pair of the second product may be written as

$[a + h] \cdot [c + k] + [b + h] \cdot [d + k]$

$= ([a + h] \cdot c + [a + h] \cdot k) + ([b + h] \cdot d + [b + h] \cdot k)$ (Why?)

$= ([a \cdot c + h \cdot c] + [a \cdot k + h \cdot k]) + ([b \cdot d + h \cdot d] + [b \cdot k + h \cdot k])$ (Why?)

$= (a \cdot c + [h \cdot c + (a \cdot k + h \cdot k)]) + (b \cdot d + [h \cdot d + (b \cdot k + h \cdot k)])$ (Why?)

$= a \cdot c + \{[h \cdot c + (a \cdot k + h \cdot k)] + (b \cdot d + [h \cdot d + (b \cdot k + h \cdot k)])\}$ (Why?)

$= a \cdot c + \{([h \cdot c + (a \cdot k + h \cdot k)] + b \cdot d) + [h \cdot d + (b \cdot k + h \cdot k)]\}$ (Why?)

$= a \cdot c + \{(b \cdot d + [h \cdot c + (a \cdot k + h \cdot k)]) + [h \cdot d + (b \cdot k + h \cdot k)]\}$ (Why?)

$= a \cdot c + \{b \cdot d + ([h \cdot c + (a \cdot k + h \cdot k)] + [h \cdot d + (b \cdot k + h \cdot k)])\}$ (Why?)

$= (a \cdot c + b \cdot d) + ([h \cdot c + (a \cdot k + h \cdot k)] + [h \cdot d + (b \cdot k + h \cdot k)])$.   (Why?)

Now, the quantity

$$([h \cdot c + (a \cdot k + h \cdot k)] + [h \cdot d + (b \cdot k + h \cdot k)])$$

involves sums and products of the natural numbers a, b, c, d, h, k and is therefore some *natural number* which we denote by r. Thus the left partner of the identifying pair of the second product is of the form

$$(a \cdot c + b \cdot d) + r.$$

It can be similarly shown (do it!) that the right partner of the identifying pair of the second product is

$$(a \cdot d + b \cdot c) + r.$$

Then, by comparing the identifying pairs of the two products,

$$(a \cdot c + b \cdot d, \ a \cdot d + b \cdot c) \quad \text{and} \quad [(a \cdot c + b \cdot d) + r, \ (a \cdot d + b \cdot c) + r],$$

we see that the partners of the second pair are simply those of the first increased by the same natural number, and therefore the pairs have the relation $\curlyvee$ and the products (integers) which they identify are equal. It is left for the reader to supply the details of the preceding argument.

Having shown that we can safely discard all except the standard representation for any integer, we recall that earlier the symbols $\ldots$, -3, -2, -1, 0, $+1$, $+2$, $+3$, $\ldots$ were introduced. It is apparent that these are also symbols for the integers, where $\ldots$, $-3 = I(0, 3)$, $-2 = I(0, 2)$, $-1 = I(0, 1)$, $0 = I(0, 0)$, $+1 = I(1, 0)$, $+2 = I(2, 0)$, $+3 = I(3, 0)$, and so on. It may seem unfortunate that zero is chosen as a symbol for the integer $I(0, 0)$ whose identifying pair has the natural number 0 as both its left and right partner, but we shall see that this will turn out to be a convenient choice. We now adopt these new symbols in place of the standard representations, calling $+1, +2, +3, +4, \ldots$ the *positive* integers and $-1, -2, -3, -4, \ldots$ the *negative* integers; 0 will be considered neither positive nor negative. Also, we shall replace the operation symbols $\oplus$ and $\otimes$ by the symbols $+$ and $\cdot$, respectively, whenever we add or multiply integers represented by our newly adopted symbols.

When we speak of the "non-negative" integers we shall mean the set

$$\{0, +1, +2, +3, +4, \ldots\}$$

and when we speak of the "non-positive" integers we shall mean the set

$$\{0, -1, -2, -3, \ldots\}.$$

There is a very obvious way of putting these two sets into $1:1$ correspondence with each other and with the natural numbers, as is illustrated below:

$$\{0, \;+1, \;+2, \;+3, \;+4, \;+5, \ldots\}$$
$$\updownarrow \quad \updownarrow \quad \updownarrow \quad \updownarrow \quad \updownarrow \quad \updownarrow$$
$$\{0, \quad 1, \quad 2, \quad 3, \quad 4, \quad 5, \ldots\}$$
$$\updownarrow \quad \updownarrow \quad \updownarrow \quad \updownarrow \quad \updownarrow \quad \updownarrow$$
$$\{0, \;-1, \;-2, \;-3, \;-4, \;-5, \ldots\}$$

When we speak of "the natural number corresponding to a given integer," we shall always mean that natural number defined by the above mapping.

We now recall some rules learned in elementary high-school algebra:

(i) to add two integers having the same sign, add their numerical values and affix their common sign, and

(ii) to add two integers having opposite signs, take the difference between their numerical values and attach the sign of the integer having the larger numerical value, except when the integers have the same numerical value, in which case their sum is 0.

Most elementary high-school algebra texts define the numerical value of an integer to be the integer "without its sign." Strictly speaking, we know that if the sign of a positive or negative integer is removed we no longer have a symbol for an integer at all, but a symbol for a *natural number*. With this in mind we now restate the two foregoing rules:

(i) to add two integers having the *same* sign, add their *corresponding natural numbers* and attach their common sign, and

(ii) to add two integers having *opposite* signs, take the difference between their *corresponding natural numbers* and attach the sign of the integer corresponding to the larger natural number, except when the integers correspond to the same natural number, in which case the sum is 0.

These rules may now be completely justified by using our original *definition* for the addition of integers: Letting h and k be any nonzero natural numbers, we have

$$(+h) + (+k) = I(h, 0) \oplus I(k, 0) = I(h + k, 0) = +(h + k),$$

and

$$(-h) + (-k) = I(0, h) \oplus I(0, k) = I(0, h + k) = -(h + k),$$

so that rule (i) is *proved*. To prove (ii) we make the following observation: if h and k are natural numbers, then

$$I(h, k) = I(h - k, k - k) = I(h - k, 0) \quad \text{if} \quad k < h,$$
$$I(h, k) = I(h - h, k - h) = I(0, k - h) \quad \text{if} \quad k > h,$$
$$I(h, k) = I(0, 0) \quad \text{if} \quad k = h.$$

Thus we can state

$$(+h) + (-k) = I(h, 0) \oplus I(0, k) = I(h, k),$$

so that

$$(+h) + (-k) = \begin{cases} +(h - k) & \text{if} \quad k < h \\ 0 & \text{if} \quad k = h \\ -(k - h) & \text{if} \quad k > h, \end{cases}$$

and (ii) is proved.

It should now be clear that one, and only one, of the integers, namely 0, is the identity element with respect to addition:

$$x + 0 = x \quad \text{for every integer} \quad x \in I.$$

This is certainly true, for if x has the representation $I(a, b)$, then

$$x + 0 = I(a, b) \oplus I(0, 0) = I(a + 0, b + 0) = I(a, b) = x.$$

We now also note that every integer has an inverse with respect to addition: an inverse of 0 is 0, an inverse of any positive integer is the corresponding negative integer, and an inverse of any negative integer is the corresponding positive integer. Therefore, since the addition of integers is associative, the system $\{I; \oplus\}$ is a group and, recalling that any element x of a group has a unique inverse, x^{-1}, we may write

$$0^{-1} = 0, \quad (+h)^{-1} = -h, \quad \text{and} \quad (-k)^{-1} = +k.$$

We have here used the symbol x^{-1} for the additive inverse of an integer x because we are familiar with this notation from our study of groups.

However, it is customary in mathematical writing to use the symbol $-x$ for the additive inverse of an integer x, and we shall conform to this convention. Thus, if h and k are nonzero natural numbers so that $+h$ is a positive integer and $-k$ is a negative integer, we write

$$-0 = 0, \quad -(+h) = -h, \quad \text{and} \quad -(-k) = +k.$$

The use of $-x$ to denote the additive inverse of an integer x may sometimes lead to confusion. For example, if the symbol -3 occurs in a certain expression, does it denote the integer -3 or the additive inverse of $+3$? Actually, either interpretation is correct, and it will usually be clear from the context which is intended.

Rules now will be stated for multiplying integers, and these rules will be justified by using the *definition* of multiplication:

(i) to multiply two integers having the same sign, multiply the corresponding natural numbers and attach a $+$ sign,

(ii) to multiply two integers having opposite signs, multiply the corresponding natural numbers and attach a $-$ sign, and

(iii) $x \cdot 0 = 0$ for any $x \in I$.

We prove (i) by taking any nonzero natural numbers h and k and observing that

$$(+h) \cdot (+k) = I(h, 0) \otimes I(k, 0) = I(h \cdot k + 0 \cdot 0, h \cdot 0 + 0 \cdot k)$$
$$= I(h \cdot k, 0) = +(h \cdot k),$$

and

$$(-h) \cdot (-k) = I(0, h) \otimes I(0, k) = I(0 \cdot 0 + h \cdot k, 0 \cdot k + h \cdot 0)$$
$$= I(h \cdot k, 0) = +(h \cdot k).$$

We prove (ii) by observing that

$$(+h) \cdot (-k) = I(h, 0) \otimes I(0, k) = I(h \cdot 0 + 0 \cdot k, h \cdot k + 0 \cdot 0)$$
$$= I(0, h \cdot k) = -(h \cdot k).$$

Finally, (iii) is proved by noting that if $I(a, b)$ is any representation for an integer x, then

$$x \cdot 0 = I(a, b) \otimes I(0, 0) = I(a \cdot 0 + b \cdot 0, a \cdot 0 + b \cdot 0) = I(0, 0) = 0.$$

The integers clearly are *not* a group with respect to multiplication, for although $+1$ is the multiplicative identity ($+1 \cdot x = x$ for all $x \in I$), not every integer has a multiplicative inverse. In fact, only two integers, $+1$ and -1, have inverses with respect to multiplication.

1. Carry out the following additions:
   (a) $[(-12) + (-7)] + (+8) =$ _____
   (b) $[(+10) + (-6)] + (-4) =$ _____
   (c) $(-3) + [(+13) + (-7)] =$ _____
   (d) $(+5) + [(-4) + (-12)] + (+8) =$ _____

2. Carry out the following indicated operations:
   (a) $(-3) \cdot [(-2) \cdot (+5)] =$ _____
   (b) $[(-4) + (+7)] \cdot [(+5) + (-8)] =$ _____
   (c) $[(-3) + (-2)] \cdot (-7) =$ _____

3. Do all parts of Problems 1 and 2 by using "standard" representation for each integer involved. If any final result is *not* a standard representation, change it to standard representation.

4. Supply all missing details in the argument on page 106.

5. On page 101 it was proved that the left partners of two identifying pairs were equal. Similarly show that the right partners are equal.

7–6 Subtraction and division. Although subtraction is not an operation on the natural numbers, it is a very simple matter to define an operation called "subtraction" on the integers because of the fact that the set of integers under addition is a group and *each integer has an additive inverse.*

DEFINITION. If x and y are any integers, we define $x - y$ (x minus y) to be that integer z such that $z = x + (-y)$.

In other words, to subtract one integer from another we add its inverse. Since every positive integer has the corresponding negative integer as its inverse and every negative integer has the corresponding positive integer as its inverse, we see that subtracting nonzero integers according to the above definition yields the same result as the rule learned in elementary algebra: "to subtract one nonzero integer from another, change the sign of the subtrahend and proceed as in addition." For example,

$$(+3) - (+5) = (+3) + (-5) = -2,$$

and

$$(+12) - (-5) = (+12) + (+5) = +17.$$

Of course, since the integer 0 is its own additive inverse, subtracting 0 yields the same as adding 0. It is important to appreciate that "subtraction" is a true operation on the integers, an operation that maps $I \times I$ into I.

Having defined the operation "subtraction" on the integers we now define the order relation "is less than" on the integers.

DEFINITION. If x and y are any integers, then $x <$ (is less than) y if, and only if, $y - x$ is a positive integer.

For example, $-3 < +5$, since $(+5) - (-3) = +8$, a positive integer, and also, since $(-3) - (-7) = +4$, a positive integer, $-7 < -3$. An easy way to tell whether one integer "is less than" another is to consider their location on the scale shown in Fig. 7–1. In this illustration, of course, we have shown only a portion of a scale extending infinitely far both left and right, with the arrow indicating the positive direction. We can say that $x < y$ if, and only if, we must traverse the scale in the positive direction to get from x to y or if, and only if, x lies to the left of y. If we wish, we can also talk about one integer being "greater than" another, where "is greater than" has the obvious meaning "is *not* less than *and* is *not* equal to."

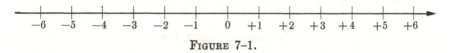

FIGURE 7–1.

Although we were able to define the operation of subtraction in terms of the more fundamental operation of addition, we are not so fortunate when it comes to division; division of one integer by another is possible only in certain circumstances. Before we formally define division we shall need the following theorem:

THEOREM. If x and y are integers such that $y \neq 0$, then there exists *at most one* integer z such that $x = y \cdot z$.

The proof of this theorem is left as an exercise. Note that the theorem does *not* say that there exists an integer z such that $x = y \cdot z$; the theorem *does* say that *if* an integer z exists such that $x = y \cdot z$, then it is the *only* such integer. We now define division:

DEFINITION. If x and y are any integers such that $y \neq 0$, we define $x \div y$ (x divided by y) to be the unique integer z, if it exists, such that $x = y \cdot z$.

It should be obvious that we cannot possibly state a suitable definition for division of an integer by 0: (i) if $x \neq 0$ then there exists no integer z such that $0 \cdot z = x$, and (ii) if $x = 0$ there exists no *unique* integer z such that $0 \cdot z = 0$. Furthermore, we see that such divisions as $(+2) \div (-5)$ and $(-7) \div (-3)$ are impossible; there exists no *integer* z such that $z \cdot (-5) = +2$ or $z \cdot (-3) = -7$. On the other hand, it is true that many such divisions as

$$(-12) \div (+4), \quad (+28) \div (-7), \quad (-15) \div (-5), \quad (+40) \div (+8)$$

are possible. We see that

$$(-12) \div (+4) = -3, \qquad (+28) \div (-7) = -4,$$
$$(-15) \div (-5) = +3, \qquad (+40) \div (+8) = +5.$$

If x and y are integers and $x \div y$ is a possible division, we call x the *dividend*, y the *divisor*, and the integer z such that $y \cdot z = x$ the *quotient*. We also say that either x "is divisible by" y or x "is a multiple of" y. For example, when we write $(-12) \div (+4) = -3$, -12 is the dividend, $+4$ is the divisor, and -3 is the quotient. In connection with the notion of divisibility we mention what are called "prime" integers. A *prime* integer is a positive integer other than $+1$ which is divisible by two, and only two, distinct positive integers, namely, itself and $+1$. Some prime integers are $+2, +3, +5, +7, +11, \ldots$. It has been known for almost 2000 years that there are infinitely many prime integers.

EXERCISE GROUP 7–3

1. (a) $(-3) + [(-7) - (-12)] =$ _____
    (b) $[(+5) - (+17)] - (-2) =$ _____
    (c) $(-4) + [(-12) \div (-3)] =$ _____
    (d) $[(-30) \div (+6)] \cdot (-5) =$ _____

2. What integers are divisors of -18?
3. What integers have $+21$ as a multiple?
4. In Problem 1 determine in which cases the brackets can be shifted.
5. Prove the theorem on page 111. [*Hint:* Assume that there are integers z and z' such that $x = y \cdot z$ and $x = y \cdot z'$ and show that $z = z'$. It will be necessary to change to standard representation.]

7–7 A simplified notation for the integers. Up to this point we have insisted that each positive integer have a $+$ sign attached to it, since we did not wish to confuse the positive integers with the natural numbers. To see if this is really necessary, let us consider the mapping β of the set N of natural numbers onto the set P_0 of non-negative integers defined schematically by

$$N = \{0, \quad 1, \quad 2, \quad 3, \quad 4, \quad 5, \ldots\},$$
$$\downarrow \quad \downarrow \quad \downarrow \quad \downarrow \quad \downarrow \quad \downarrow$$
$$P_0 = \{0, +1, +2, +3, +4, +5, \ldots\}.$$

We observe that $\beta n = +n \in P_0$ for each $n \in N$ ($n \neq 0$) and, for $0 \in N$, we have $\beta 0 = 0 \in P_0$. Also, we see that for all $a, b \in N$,

$$\beta(a + b) = (\beta a) + (\beta b) \qquad \text{and} \qquad \beta(a \cdot b) = (\beta a) \cdot (\beta b).$$

Furthermore, if $a, b \in N$ and $a < b$, then $\beta a, \beta b \in P_0$ and $\beta a < \beta b$; the "is less than" relation is preserved by the mapping β. The mapping $\beta : N \rightarrow P_0$ is $1 : 1$ and is clearly an isomorphism of $\{N; +\}$ onto $\{P_0; +\}$ and also of $\{N; \cdot\}$ onto $\{P_0; \cdot\}$. In short, the number systems $\{N; +, \cdot\}$ and $\{P_0; +, \cdot\}$ have *the same structure*. This means, as the reader is probably aware, that any problem requiring only the application of natural number arithmetic for its solution may be solved just as satisfactorily by using the arithmetic of non-negative integers. Because the system $\{P_0; +, \cdot\}$ may be used in place of the system $\{N; +, \cdot\}$, we habitually make no distinction between the natural numbers and the non-negative integers; we have come to think of these systems as being the same. Thus, since we ordinarily make no distinction between the natural numbers 0, 1, 2, 3, . . . and the non-negative integers 0, +1, +2, +3, . . . , we simply use the symbols 0, 1, 2, 3, 4, . . . for both.

To those not familiar with the process we employed in constructing the system $\{I; \oplus, \otimes\}$, it would appear that we have merely attached the list of symbols

$$\ldots, -4, -3, -2, -1$$

to the list

$$0, 1, 2, 3, 4, \ldots$$

of natural numbers and have defined both addition and multiplication on this enlarged set so as to be consistent with (agree with) the operations $(+)$ and $(\cdot)$ on the natural numbers. Of course, this is *not* the way we constructed the system $\{I; \oplus, \otimes\}$, but since it *appears to be*, we are in the habit of saying that we have "extended" the natural numbers and that the natural numbers are "embedded" in the integers (in other words, the set N is a subset of the set I).

EXERCISE GROUP 7–4

1. Determine whether or not the natural numbers and the nonpositive integers are isomorphic with respect to addition; with respect to multiplication.

2. Could the nonpositive integers be used for counting, addition, and multiplication? Explain.

7–8 Integral domains. We have studied number systems, where a number system is a nonempty set S together with two operations, $\oplus$ and $\odot$, on S, both operations being commutative and associative and $\odot$ being distributive with respect to $\oplus$. We have called $\oplus$ and $\odot$ "addition" and "multiplication," respectively, although this in no sense implied that $\oplus$ and $\odot$ were ordinary addition and multiplication. We are now going to study a *special kind* of number system called an *integral domain* and give examples of such a system.

DEFINITION. A number system $\{S; \oplus, \odot\}$ is an integral domain if, and only if, it possesses each of the following properties:

(i) S is a group with respect to $\oplus$,
(ii) S contains unique and distinct elements z and u such that $z \oplus a = a$ and $u \odot a = a$ for each element $a \in S$, and
(iii) if $a, b, c \in S$ and $a \neq z$, then $a \odot b = a \odot c$ implies that $b = c$.

An immediate example of an integral domain is the number system $\{I; \oplus, \otimes\}$. We note that I is a group with respect to $\oplus$ and that there are unique and distinct elements 0 and 1 in I such that $0 + x = x$ and $1 \cdot x = x$ for each $x \in I$. To show that $\{I; \oplus, \otimes\}$ possesses Property (iii), we take any integers x, y, z such that $x \neq 0$ and $x \cdot y = x \cdot z$. Since $x \neq 0$, $x = +h$ or $-h$ for some natural number $h \neq 0$ and, for definiteness, we suppose that $x = +h$. Then, using the representations $I(h, 0)$, $I(a, b)$, $I(c, d)$ for x, y, z, respectively, we may write

$$I(h, 0) \otimes I(a, b) = I(h, 0) \otimes I(c, d),$$
$$I(h \cdot a + 0 \cdot b, h \cdot b + 0 \cdot a) = I(h \cdot c + 0 \cdot d, h \cdot d + 0 \cdot c),$$
$$I(h \cdot a, h \cdot b) = I(h \cdot c, h \cdot d),$$

where a, b, c, d are natural numbers. Now, since

$$I(h \cdot a, h \cdot b) = I(h \cdot c, h \cdot d),$$

the identifying pairs $(h \cdot a, h \cdot b)$, $(h \cdot c, h \cdot d)$ have the relation $\curvearrowright$, so that

$$h \cdot a + h \cdot d = h \cdot b + h \cdot c,$$

and since multiplication of natural numbers is distributive with respect to addition (Axiom IV, Section 2–2), we may write

$$h \cdot (a + d) = h \cdot (b + c).$$

Because $h \neq 0$ we can use the cancellation property for multiplication (Axiom V) and write $a + d = b + c$ which, in turn, means that $(a, b) \curvearrowright (c, d)$. Thus $I(a, b) = I(c, d)$, since their identifying pairs have the relation $\curvearrowright$, and since $I(a, b)$, $I(c, d)$ are representations for y and z, it follows that $y = z$, as we wished to show; we have proved the cancellation property for multiplication. Therefore $\{I; \oplus, \odot\}$ possesses the properties of an integral domain.

Note: $x, y, z, \ldots$ represent integers.

1. If $x + y = x + z$, prove that $y = z$. (This is the cancellation property for addition.) [*Hint:* Let x, y, z be represented by $I(a, b)$, $I(c, d)$, $I(e, f)$, respectively, and prove that $(c, d) \smallfrown (e, f)$.]

2. If $x = y$, prove that $x + z = y + z$. (Use the substitution principle!)

3. If $w = x$ and $y = z$, prove that $w + y = x + z$. (Use the result of Problem 2.)

4. If $x = y$, prove that $x - z = y - z$. [Remember: We *define* $x - z = x + (-z)$.]

5. Show that $-(-x) = x$.

6. Show that $-x = (-1) \cdot x$.

7. Show that $x - (y - z) = (x - y) + z$.

8. If $x < y$, prove that $x + z < y + z$. [*Hint:* Show that $(y + z) - (x + z)$ is a positive integer; that is, that $0 < (y + z) - (x + z)$.]

9. If $x < y$ and $0 < z$, prove that $x \cdot z < y \cdot z$.

10. If $x < y$ and $z < 0$, prove that $x \cdot z > y \cdot z$.

7-9 Congruences. It is our purpose in this section to obtain further examples of integral domains and we start out by discussing a relation on the integers, the relation $\equiv_m$, read "is congruent modulo m to," where m is an integer. We define this relation as follows:

DEFINITION. If a and b are any integers, then $a \equiv_m b$ if, and only if, there exists an integer k such that $a - b = m \cdot k$.

For example, $7 \equiv_4 15$ because $7 - 15 = (-2) \cdot 4$; $12 \equiv_5 -3$ because $12 - (-3) = 3 \cdot 5$; and $-14 \equiv_8 2$ because $-14 - 2 = (-2) \cdot 8$. An interesting thing about the relation $\equiv_m$ is that it is an equivalence relation on the set of integers, and we can easily show this:

(i) For any $x \in I$ it is true that $x \equiv_m x$, for $x - x = 0 \cdot m$. The relation $\equiv_m$ is reflexive.

(ii) If $x, y \in I$ and $x \equiv_m y$, there exists an integer k such that $x - y = k \cdot m$. Therefore, $y - x = -(k \cdot m) = (-k) \cdot m$ and $y \equiv_m x$, so that the relation $\equiv_m$ is symmetric.

(iii) If $x, y, z \in I$, $x \equiv_m y$ and $y \equiv_m z$, there exist integers h and k such that

$$x - y = h \cdot m \qquad \text{and} \qquad y - z = k \cdot m.$$

Therefore

$$x - z = x - y + y - z = (x - y) + (y - z) = h \cdot m + k \cdot m = (h + k) \cdot m$$

and $x \equiv_m z$, so that the relation $\equiv_m$ is transitive.

Now that we know $\equiv_m$ to be an equivalence relation on I, we can use this relation to induce a partition of I. We choose a particular value of m, say $m = 3$. For each integer $x \in I$ let I_x denote the set of all $y \in I$ such that $y \equiv_3 x$. We see that

$$I_0 = \{\ldots, -6, -3, 0, 3, 6, 9, \ldots\},$$
$$I_1 = \{\ldots, -5, -2, 1, 4, 7, 10, \ldots\},$$
$$I_2 = \{\ldots, -4, -1, 2, 5, 8, \ldots\},$$
$$I_{-1} = \{\ldots, -4, -1, 2, 5, 8, \ldots\},$$

and so on. We see that we can construct infinitely many of these equivalence sets, but we also see that

$$\cdots = I_{-6} = I_{-3} = I_0 = I_3 = I_6 = \cdots,$$
$$\cdots = I_{-5} = I_{-2} = I_1 = I_4 = I_7 = \cdots,$$
$$\cdots = I_{-4} = I_{-1} = I_2 = I_5 = I_8 = \cdots,$$

so that the distinct and disjoint sets I_0, I_1, I_2 give us our partition of I. It is our purpose now to define operations $\oplus$ and $\odot$ on the family of equivalence sets $\{I_0, I_1, I_2\}$ in such a way that

$$\{\{I_0, I_1, I_2\}; \oplus, \odot\}$$

will be a number system. However, to guarantee in advance that the operations $\oplus$ and $\odot$, as we are about to define them, will yield unique sums and products, we shall need a theorem. In stating the theorem, which follows, the symbol K will denote the set $\{0, 1, 2\}$ of integers.

THEOREM. For any $i, j \in K$ there is a unique integer $h \in K$ such that $h \equiv_3 i + j$ and there is a unique integer $k \in K$ such that $k \equiv_3 i \cdot j$.

To prove this theorem we must show (i) that there are integers $h, k \in K$ such that $h \equiv_3 i + j$ and $k \equiv_3 i \cdot j$, and (ii) that they are unique. To prove (i) we simply examine the following tables, remembering that addition and multiplication of integers are commutative:

$0 + 0 = 0 \equiv_3 0$		$0 \cdot 0 = 0 \equiv_3 0$
$0 + 1 = 1 \equiv_3 1$		$0 \cdot 1 = 0 \equiv_3 0$
$0 + 2 = 2 \equiv_3 2$		$0 \cdot 2 = 0 \equiv_3 0$
$1 + 1 = 2 \equiv_3 2$	and	$1 \cdot 1 = 1 \equiv_3 1$
$1 + 2 = 3 \equiv_3 0$		$1 \cdot 2 = 2 \equiv_3 2$
$2 + 2 = 4 \equiv_3 1$		$2 \cdot 2 = 4 \equiv_3 1$

It is clear from these tables that for any $i, j \in K$ there is an integer (at least one) $h \in K$ such that $h \equiv_3 i + j$ and there is an integer (at least one) $k \in K$ such that $k \equiv_3 i \cdot j$. To show (ii) we suppose that there are integers $h', k' \in K$ such that $h' \equiv_3 i + j$ and $k' \equiv_3 i \cdot j$. Hence, since $\equiv_3$ is reflexive and transitive, $h' \equiv_3 h$ and $k' \equiv_3 k$. Next, we recall that $h' \equiv_3 h$ means that $h' - h = p \cdot 3$ for some integer p, or what is the same thing, that $h' - h$ is divisible by 3. Thus, since $h', h \in K$, $h' - h$ is one of the integers $0, +1, -1, +2, -2$, and since 0 is the only one of these integers divisible by 3, it follows that $h' - h = 0$ or $h' = h$. A similar argument will show that $k' = k$, and we will be assured that both h and k are unique integers. With the theorem proved, we are now in position to define $\oplus$ (addition) and $\odot$ (multiplication) on the set $\{I_0, I_1, I_2\}$.

Addition. We define $I_i \oplus I_j = I_k$, where $k \in K$ and $k \equiv_3 i + j$. Using this definition, we can construct the following table.

$\oplus$	I_0	I_1	I_2
I_0	I_0	I_1	I_2
I_1	I_1	I_2	I_0
I_2	I_2	I_0	I_1

The symmetry of the table tells us at once that $\oplus$ is commutative. We note also that I_0 is the unique additive identity and that each element has an additive inverse: I_0 is an inverse of I_0, I_2 is an inverse of I_1, and I_1 is an inverse of I_2. To show that $\{I_0, I_1, I_2\}$ is a group with respect to $\oplus$, it remains only to show that $\oplus$ is an associative operation. To do this, we must first prove the following *lemma* (preliminary theorem):

LEMMA. *If x and y are integers and $x \equiv_m y$, then $x + z \equiv_m y + z$ for any integer z.*

Proof. We observe that $x \equiv_m y$ implies that $x - y = s \cdot m$ for some integer s. Then it is also true that

$$(x - y) + 0 = (x - y) + (z - z) = (x + z) - (y + z) = s \cdot m$$

for any integer z, which in turn implies that $x + z \equiv_m y + z$ for any integer z.

We now consider

$$I_h \oplus (I_i \oplus I_j) \quad \text{and} \quad (I_h \oplus I_i) \oplus I_j.$$

If $k \in K$ and $k \equiv_3 i + j$, we may write

$$I_h \oplus (I_i \oplus I_j) = I_h \oplus I_k.$$

Also, if $g \in K$ and $g \equiv_3 h + i$, we may write

$$(I_h \oplus I_i) \oplus I_j = I_g \oplus I_j.$$

Our problem now is that of showing $I_h \oplus I_k = I_g \oplus I_j$. Applying the foregoing lemma, we see that

$$i + j \equiv_3 k \qquad \text{implies} \qquad h + (i + j) \equiv_3 h + k$$

and

$$h + i \equiv_3 g \qquad \text{implies} \qquad (h + i) + j \equiv_3 g + j.$$

Therefore, since $\equiv_3$ is transitive, $h + k \equiv_3 g + j$. Finally, we let

$$I_h \oplus I_k = I_p \qquad \text{and} \qquad I_g \oplus I_j = I_q,$$

where $p, q \in K$, $p \equiv_3 h + k$, and $q \equiv_3 g + j$. Since any two elements of K are $\equiv_3$ if and only if they are equal, it follows that $p = q$ and $I_p = I_q$. We have thus proved that $\oplus$ is associative and that $\{I_0, I_1, I_2\}$ is a group with respect to $\oplus$.

As a matter of interest, let us observe the obvious isomorphism between the system

$$\{\{I_0, I_1, I_2\}; \oplus\}$$

and the system $\{S; \oplus\}$ on page 69:

$$0 \leftrightarrow I_0, \qquad 1 \leftrightarrow I_1, \qquad 2 \leftrightarrow I_2.$$

We can, if we wish, replace one system with the other, for they have the same structure, i.e., they are mathematically equivalent.

Multiplication. We define $I_i \odot I_j = I_k$, where $k \in K$ and $k \equiv_3 i \cdot j$. Application of this definition yields the following table.

$\odot$	I_0	I_1	I_2
I_0	I_0	I_0	I_0
I_1	I_0	I_1	I_2
I_2	I_0	I_2	I_1

The symmetry of the table tells us that $\odot$ is commutative and the table also shows that I_1 is the unique multiplicative identity. To show that $\odot$

is associative, we consider

$$I_h \odot (I_i \odot I_j) \quad \text{and} \quad (I_h \odot I_i) \odot I_j.$$

If $k \in K$, $k \equiv_3 i \cdot j$ and $g \in K$, $g \equiv_3 h \cdot i$, we may write

$$I_h \odot (I_i \odot I_j) = I_h \odot I_k \quad \text{and} \quad (I_h \odot I_i) \odot I_j = I_g \odot I_j,$$

so that proving associativity reduces to proving

$$I_h \odot I_k = I_g \odot I_j.$$

We shall need the following lemma:

LEMMA. If x, y are integers and $x \equiv_m y$, then $x \cdot z \equiv_m y \cdot z$ for any integer z.

Proof. Since $x \equiv_m y$, it is true that $x - y = k \cdot m$ for some integer k. Thus for any integer z,

$$(x - y) \cdot z = (k \cdot m) \cdot z.$$

Using the distributive law on the left and the associative and commutative laws on the right, we may write

$$(x \cdot z) - (y \cdot z) = (k \cdot z) \cdot m,$$

which proves $x \cdot z \equiv_m y \cdot z$.

Now, appealing to this lemma, we have

$$k \equiv_3 i \cdot j \quad \text{and} \quad g \equiv_3 h \cdot i,$$

so that

$$h \cdot k \equiv_3 h \cdot (i \cdot j) \quad \text{and} \quad g \cdot j \equiv_3 (h \cdot i) \cdot j.$$

Therefore, since $\equiv_3$ is transitive, $h \cdot k \equiv_3 g \cdot j$. Now, if $p \in K$, $p \equiv_3 h \cdot k$, and if $q \in K$, $q \equiv_3 g \cdot j$, we have

$$I_h \odot I_k = I_p \quad \text{and} \quad I_g \odot I_j = I_q.$$

Since $p, q \in K$ and $p \equiv_3 q$, it follows that $p = q$ and $I_p = I_q$. We have thus proved $\odot$ to be associative.

To prove that $\odot$ is distributive with respect to $\oplus$, we consider

$$I_g \odot (I_h \oplus I_i) \quad \text{and} \quad (I_g \odot I_h) \oplus (I_g \odot I_i).$$

Let

$$I_g \odot (I_h \oplus I_i) = I_g \odot I_j,$$

where $j \in K$ and $j \equiv_3 h + i$; let

$$I_g \odot I_j = I_k,$$

where $k \in K$ and $k \equiv_3 g \cdot j$; let

$$I_g \odot I_h = I_p,$$

where $p \equiv_3 g \cdot h$; let

$$I_g \odot I_i = I_q,$$

where $q \in K$ and $q \equiv_3 g \cdot i$; and, finally, let

$$I_p \oplus I_q = I_r,$$

where $r \in K$ and $r \equiv_3 p + q$. We must now prove that $k = r$, and we will use both of the lemmas we have proved. Since $h + i \equiv_3 j$, then

$$g \cdot (h + i) \equiv_3 g \cdot j,$$

and since $g \cdot j \equiv_3 k$, then

$$g \cdot (h + i) \equiv_3 k \quad \text{(Why?)}.$$

Since $g \cdot h \equiv_3 p$, then

$$(g \cdot h) + (g \cdot i) \equiv_3 p + (g \cdot i),$$

and since $g \cdot i \equiv_3 q$, then

$$p + (g \cdot i) \equiv_3 p + q.$$

Hence, by transitivity,

$$(g \cdot h) + (g \cdot i) \equiv_3 p + q,$$

and, since $p + q \equiv_3 r$,

$$(g \cdot h) + (g \cdot i) \equiv_3 r.$$

Now, by the distributive law,

$$g \cdot (h + i) = (g \cdot h) + (g \cdot i)$$

and, again by transitivity, $k \equiv_3 r$. However, since k and r are integers from $K = \{0, 1, 2\}$, it follows that $k = r$ and the operation $\odot$ is distributive, as we wished to show.

So far we have shown that

$$\{\{I_0, I_1, I_2\}; \oplus, \odot\}$$

is a number system, $\{I_0, I_1, I_2\}$ is a group with respect to $\oplus$, and there are distinct additive and multiplicative identity elements in the set $\{I_0, I_1, I_2\}$. To show that the system

$$\{\{I_0, I_1, I_2\}; \oplus, \odot\}$$

is an integral domain, we must finally prove that it possesses the multiplicative cancellation property; that is, we must show that if

$$I_g \odot I_h = I_g \odot I_i \quad \text{and} \quad I_g \neq I_0,$$

then $I_h = I_i$. To show this, we first observe that

$$I_g \odot I_h = I_g \odot I_i$$

implies that

$$g \cdot h \equiv_3 j \quad \text{and} \quad g \cdot i \equiv_3 j$$

for some integer $j \in K$. Therefore, by transitivity, $g \cdot h \equiv_3 g \cdot i$. This congruence, of course, holds true if $h = i$, so we will show that it cannot possibly hold true if $h \neq i$:

if $g = 1, h = 0$, and $i = 1$, then $1 \cdot 0 \not\equiv_3 1 \cdot 1$,

if $g = 1, h = 0$, and $i = 2$, then $1 \cdot 0 \not\equiv_3 1 \cdot 2$,

if $g = 1, h = 1$, and $i = 2$, then $1 \cdot 1 \not\equiv_3 1 \cdot 2$,

if $g = 2, h = 0$, and $i = 1$, then $2 \cdot 0 \not\equiv_3 2 \cdot 1$,

if $g = 2, h = 0$, and $i = 2$, then $2 \cdot 0 \not\equiv_3 2 \cdot 2$,

if $g = 2, h = 1$, and $i = 2$, then $2 \cdot 1 \not\equiv_3 2 \cdot 2$.

Interchanging the values of h and i will yield similar results, and since all cases in which $h \neq i$ yield $g \cdot h \not\equiv_3 g \cdot i$, the fact that $g \cdot h \equiv_3 g \cdot i$ must imply that $h = i$, so that $I_h = I_i$. This proves that the number system

$$\{\{I_0, I_1, I_2\}; \oplus, \odot\}$$

is an integral domain.

EXERCISE GROUP 7–6

1. In proving the first of the two lemmas in Section 7–9 the statement

$$(x - y) + (z - z) = (x + z) - (y + z)$$

was made. Prove this statement.

2. In proving that $\equiv_m$ is a symmetric relation the statement $-(k \cdot m) = (-k) \cdot m$ was made. Prove this statement.

3. In proving that $\equiv_m$ is a transitive relation the statement

$$x - z = (x - y) + (y - z)$$

was made. Prove this statement.

4. Use the equivalence relation $\equiv_4$ to partition the integers and use the subsets I_0, I_1, I_2, I_3 to construct a number system, where addition and multiplication are defined as follows:

Addition: $I_i \oplus I_j = I_k$, where $k \in \{0, 1, 2, 3\}$ and $k \equiv_4 i + j$,

Multiplication: $I_i \odot I_j = I_k$, where $k \in \{0, 1, 2, 3\}$ and $k \equiv_4 i \cdot j$.

Construct the $\oplus$ and $\odot$ tables. Show that

$$\big\{ \{I_0, I_1, I_2, I_3\}; \oplus, \odot \big\}$$

is *not* an integral domain. Explain.

5. Partition the integers, using the equivalence relation $\equiv_0$; using the equivalence relation $\equiv_1$.

6. Consider the partition $\{I_0, I_1, I_2\}$ of I induced by the equivalence relation $\equiv_3$ in Section 7–9. Is the subset $\{I_1, I_2\}$ a group with respect to $\odot$?

7–10 Conclusion. The system $\{I; \oplus, \otimes\}$ possesses an obvious advantage over the natural number system; each integer has an inverse with respect to addition and, of course, this is what makes subtraction possible. However, division cannot be called an operation on the set I; division of one integer by another is possible only in special cases. It is for this reason that the next chapter will be devoted to the development of the *rational number system.* The rational number system will possess all the properties of the integers and, in addition, every rational number except the zero element will have a multiplicative inverse. Thus division, except by zero, will be possible. We shall also show that the integers are isomorphic to a certain subset of the rationals with respect to both addition and multiplication and we will be able to think of the rationals as being an extension of the integers—just as we may think of the integers as being an extension of the natural numbers.

CHAPTER 8

THE RATIONAL NUMBERS

8-1 Constructing the rationals. In Section 7–10 we stated our intention of constructing a new number system, the *rational* number system. We will start in the usual way by first constructing a set. Later we will define operations on the set and show that we indeed have a number system. To begin, then, we take the set

$$I = \{\ldots, \; -3, \; -2, \; -1, \; 0, \; 1, \; 2, \; 3, \; 4, \ldots\}$$

of integers and the set

$$J = \{\ldots, \; -3, \; -2, \; -1, \; 1, \; 2, \; 3, \; 4, \ldots\}$$

of nonzero integers and construct the set $I \times J$ of all ordered pairs of integers whose right partners are not zero. It is important to keep in mind that if $(x, y) \in I \times J$, then $y \neq 0$. Next, we define the relation $\backsim$ (wiggle) on the set $I \times J$ as follows.

DEFINITION. If $(r, s), (t, u) \in I \times J$, then $(r, s) \backsim (t, u)$ if, and only if, $r \cdot u = s \cdot t$.

When we see the symbol $\backsim$ between two ordered pairs of integers we will read "has the wiggle relation to." Thus, $(2, 4) \backsim (3, 6)$ would be read "$(2, 4)$ has the wiggle relation to $(3, 6)$" since $2 \cdot 6 = 4 \cdot 3$. We should keep in mind that the wiggle relation $\backsim$ is a subset of $(I \times J) \times (I \times J)$, each of whose members is an ordered pair whose left and right partners are ordered pairs of integers from $I \times J$. For example, the pairs $[(2, 4), (3, 6)]$ and $[(-5, 12), (10, -24)]$ would be found in the relation $\backsim$. You have probably guessed by now that $\backsim$ is an equivalence relation, and we proceed to verify this.

(i) For each $(r, s) \in I \times J$, $(r, s) \backsim (r, s)$, for $r \cdot s = s \cdot r$. Therefore, $\backsim$ is a reflexive relation.

(ii) If $(r, s), (t, u) \in I \times J$ and $(r, s) \backsim (t, u)$, then $r \cdot u = s \cdot t$. But this implies that $t \cdot s = u \cdot r$, so that $(t, u) \backsim (r, s)$. Thus $\backsim$ is a symmetric relation.

(iii) If $(r, s), (t, u), (v, w) \in I \times J$, $(r, s) \backsim (t, u)$, and $(t, u) \backsim (v, w)$, then $r \cdot u = s \cdot t$ and $t \cdot w = u \cdot v$. Thus

$$(r \cdot u) \cdot w = (s \cdot t) \cdot w \qquad \text{and} \qquad s \cdot (t \cdot w) = s \cdot (u \cdot v),$$

so that $(r \cdot u) \cdot w = s \cdot (u \cdot v)$. By using the commutative and associative laws we may therefore write

$$u \cdot (r \cdot w) = u \cdot (s \cdot v),$$

and since $u \neq 0$, $r \cdot w = s \cdot v$. Therefore $(r, s) \backsim (v, w)$ and $\backsim$ is a transitive relation.

Thus $\backsim$ is reflexive, symmetric, and transitive and is an equivalence relation on the set $I \times J$. We now take the obvious step and use the relation $\backsim$ to partition the set $I \times J$ as follows: For each $(r, s) \in I \times J$ let $R(r, s)$ (read "capital R of r and s") be the set of all elements $(t, u) \in I \times J$ such that $(t, u) \backsim (r, s)$. For example, the set $R(2, 3)$ is the set of all pairs $(t, u) \in I \times J$ such that $(t, u) \backsim (2, 3)$. Some of the members of $R(2, 3)$ are $(2, 3)$, $(-2, -3)$, and $(6, 9)$. One thing we know with certainty: The partition of $I \times J$ into subsets of the form $R(r, s)$ is a partition induced by an equivalence relation, and any sets $R(r, s)$ and $R(t, u)$ are either the same or disjoint; if two subsets of the partition have *one* member in common they have *all* members in common. This makes us realize that any particular subset of the partition has infinitely many symbolic representations. For example, $R(2, 3)$, $R(-2, -3)$, $R(4, 6)$, $R(-6, -9)$ are merely different symbols for the same set. We will be able to show later that we have complete freedom in choosing a symbol to represent a particular set. We shall call the subsets of this partition *rationals*.

EXERCISE GROUP 8–1

1. Write five other representations for the rational which has $R(-2, 5)$ as one of its representations.

2. Prove that $R(3, 4) = R(9, 12)$ by showing that $R(3, 4) \subset R(9, 12)$ and $R(9, 12) \subset R(3, 4)$. Make your argument complete.

8–2 The operations ⊞ and ⊠ on the rationals. We shall define addition of rationals as follows:

DEFINITION. For any rationals $R(s, t)$ and $R(u, v)$ we define

$$R(s, t) \boxplus R(u, v) = R(s \cdot v + t \cdot u, \ t \cdot v),$$

where $R(s \cdot v + t \cdot u, \ t \cdot v)$ is a rational, since $t \cdot v \neq 0$ and

$$(s \cdot v + t \cdot u, \ t \cdot v) \in I \times J.$$

Before we commit ourselves to final adoption of this definition we must show that addition has been "well defined"; that is, that the "sum"

$$R(s \cdot v + t \cdot u, \, t \cdot v)$$

is independent of the representation used for the rationals that were added. To show this, suppose that $R(w, x)$ and $R(y, z)$ are different representations for $R(s, t)$ and $R(u, v)$, respectively, so that

$$(w, x) \cap (s, t) \qquad \text{and} \qquad (y, z) \cap (u, v).$$

Then

$$R(w, x) \; \boxed{+} \; R(y, z) \; = \; R(w \cdot z + x \cdot y, \, x \cdot z),$$

and we assert that

$$(w \cdot z + x \cdot y, \, x \cdot z) \cap (s \cdot v + t \cdot u, \, t \cdot v),$$

so that

$$R(w \cdot z + x \cdot y, \, x \cdot z) = R(s \cdot v + t \cdot u, \, t \cdot v).$$

Proof. Since $(w, x) \cap (s, t)$, $w \cdot t = x \cdot s$, and since $(y, z) \cap (u, v)$, $y \cdot v = z \cdot u$. We now multiply both sides of $w \cdot t = x \cdot s$ by $z \cdot v \neq 0$ and multiply both sides of $y \cdot v = z \cdot u$ by $x \cdot t \neq 0$, thus obtaining

$$(w \cdot t) \cdot (z \cdot v) = (x \cdot s) \cdot (z \cdot v) \quad \text{and} \quad (y \cdot v) \cdot (x \cdot t) = (z \cdot u) \cdot (x \cdot t).$$

Addition of these results yields

$$(w \cdot t) \cdot (z \cdot v) + (y \cdot v) \cdot (x \cdot t) = (x \cdot s) \cdot (z \cdot v) + (z \cdot u) \cdot (x \cdot t),$$

and, by applying the commutative and associative laws for multiplication of integers, we obtain

$$(w \cdot z) \cdot (t \cdot v) + (x \cdot y) \cdot (t \cdot v) = (s \cdot v) \cdot (x \cdot z) + (t \cdot u) \cdot (x \cdot z).$$

Finally, then, by the distributive law, we have

$$(w \cdot z + x \cdot y) \cdot (t \cdot v) = (s \cdot v + t \cdot u) \cdot (x \cdot z),$$

which means that

$$(w \cdot z + x \cdot y, \, x \cdot z) \cap (s \cdot v + t \cdot u, \, t \cdot v),$$

and

$$R(w \cdot z + x \cdot y, \, x \cdot z) = R(s \cdot v + t \cdot u, \, t \cdot v).$$

Thus when adding two rationals we can choose any representations we please for them and be sure that we will obtain a unique sum.

We shall define multiplication of rationals as follows.

DEFINITION. For any rationals $R(s, t)$ and $R(u, v)$ we define

$$R(s, t) \boxed{\times} R(u, v) = R(s \cdot u, t \cdot v).$$

Again, we must show that multiplication has been "well defined." To show that our result (product) is unique, let $R(w, x)$ and $R(y, z)$ be other representations for $R(s, t)$ and $R(u, v)$, respectively, where

$$(w, x) \cap (s, t) \qquad \text{and} \qquad (y, z) \cap (u, v),$$

so that

$$w \cdot t = x \cdot s \qquad \text{and} \qquad y \cdot v = z \cdot u.$$

Upon multiplying these results we obtain

$$(w \cdot t) \cdot (y \cdot v) = (x \cdot s) \cdot (z \cdot u),$$

so that, by the commutative and associative laws, we have

$$(w \cdot y) \cdot (t \cdot v) = (x \cdot z) \cdot (s \cdot u),$$

which implies that

$$(w \cdot y, x \cdot z) \cap (s \cdot u, t \cdot v) \quad \text{and} \quad R(w \cdot y, x \cdot z) = R(s \cdot u, t \cdot v).$$

Thus when multiplying any two rationals we can choose any representations we please for them and be certain that our product will be unique.

EXERCISE GROUP 8–2

1. Carry out the additions

$$[R(2, -1) \boxed{+} R(3, 5)] \boxed{+} R(5, 3)$$

and

$$R(2, -1) \boxed{+} [R(3, 5) \boxed{+} R(5, 3)].$$

Show that the sums obtained are equal.

2. Show that $R(5, 2) \boxed{\times} R(-1, 3) = R(10, 4) \boxed{\times} R(3, -9)$.

8–3 The commutative and associative laws. The fact that both addition and multiplication are commutative and associative operations on the rationals is a direct consequence of the commutativity and associativity of both addition and multiplication on the integers from which the rationals are descended; these properties seem indeed to be *inherited* properties. The *commutative law for addition* of rationals is proved by simply noting that

$$R(s, t) \boxplus R(u, v) = R(s \cdot v + t \cdot u, t \cdot v)$$
$$= R(u \cdot t + v \cdot s, v \cdot t)$$
$$= R(u, v) \boxplus R(s, t).$$

To prove the *associative law for addition* of rationals observe that

$$R(s, t) \boxplus [R(u, v) \boxplus R(w, x)] = R(s, t) \boxplus R(u \cdot x + v \cdot w, v \cdot x)$$
$$= R[s \cdot (v \cdot x) + t \cdot (u \cdot x + v \cdot w), t \cdot (v \cdot x)]$$
$$= R[(s \cdot v + t \cdot u) \cdot x + (t \cdot v) \cdot w, (t \cdot v) \cdot x]$$
$$= R(s \cdot v + t \cdot u, t \cdot v) \boxplus R(w, x)$$
$$= [R(s, t) \boxplus R(u, v)] \boxplus R(w, x).$$

We now ask if there is any rational which plays the role of an additive identity, and it is easily seen that any rational of the form $R(0, u)$ plays such a role. First we observe that if

$$(u, v) \in I \times J \qquad \text{and} \qquad s \neq 0,$$

then $(u \cdot s, v \cdot s) \in I \times J$ also and, moreover, $(u, v) \curlyvee (u \cdot s, v \cdot s)$, since $u \cdot (v \cdot s) = v \cdot (u \cdot s)$. Therefore, if $R(s, t)$ is any rational,

$$R(0, u) \boxplus R(s, t) = R(0 \cdot t + u \cdot s, u \cdot t)$$
$$= R(u \cdot s, u \cdot t)$$
$$= R(s, t), \quad \text{since} \quad (s, t) \curlyvee (u \cdot s, u \cdot t).$$

Now, if u and v are any nonzero integers, it is true that $(0, u) \curlyvee (0, v)$ and it follows that

$$\ldots, \; R(0, -2), \; R(0, -1), \; R(0, 1), \; R(0, 2), \ldots$$

are all representations for the unique rational which plays the role of the additive identity. At this point we should also note that each rational $R(u, v)$ has an additive inverse, namely, $R(-u, v)$, where $-u$ denotes the additive inverse (negative) of the integer u. We observe that

$$R(u, v) \boxplus R(-u, v) = R(u \cdot v + v \cdot (-u), v^2)$$
$$= R(v \cdot [u + (-u)], v^2)$$
$$= R(v \cdot 0, v^2)$$
$$= R(0, v^2), \quad \text{the additive identity.}$$

We customarily denote the additive inverse of any rational $R(u, v)$ by $-R(u, v)$ just as we denote the additive inverse of any integer u by $-u$;

in other words, we *define* $-R(u, v) = R(-u, v)$. It is now clear that the rationals are a group with respect to addition, for (i) addition is associative, (ii) there exists a unique identity element, and (iii) each rational has an additive inverse.

The commutative and associative laws for multiplication of rationals are quite easy to prove. Observe that

$$R(s, t) \boxtimes R(u, v) = R(s \cdot u, t \cdot v)$$
$$= R(u \cdot s, v \cdot t)$$
$$= R(u, v) \boxtimes R(s, t),$$

which proves the commutative law. Also observe that

$$R(s, t) \boxtimes [R(u, v) \boxtimes R(w, x)] = R(s, t) \boxtimes R(u \cdot w, v \cdot x)$$
$$= R(s \cdot [u \cdot w], t \cdot [v \cdot x])$$
$$= R([s \cdot u] \cdot w, [t \cdot v] \cdot x)$$
$$= R(s \cdot u, t \cdot v) \boxtimes R(w, x)$$
$$= [R(s, t) \boxtimes R(u, v)] \boxtimes R(w, x),$$

which proves the associative law. Both of these laws obviously hold because multiplication of integers is both commutative and associative. The fact that multiplication on the rationals is distributive is not difficult to prove, but is tedious. We must show that

$$R(s, t) \boxtimes [R(u, v) \boxplus R(w, x)] = R(s, t) \boxtimes R(u, v) \boxplus R(s, t) \boxtimes R(w, x).$$

Working with the left side of this equality gives us

$$R(s, t) \boxtimes R(u \cdot x + v \cdot w, v \cdot x) = R(s \cdot [u \cdot x + v \cdot w], t \cdot [v \cdot x]),$$

and working with the right side yields

$$R(s \cdot u, t \cdot v) \boxplus R(s \cdot w, t \cdot x) = R([s \cdot u] \cdot [t \cdot x] + [t \cdot v] \cdot [s \cdot w], [t \cdot v] \cdot [t \cdot x]).$$

To show that these rationals are equal we need only prove that the identifying pair of one rational has the relation $\curlywedge$ to the identifying pair of the other. By using the commutative, associative, and distributive laws, we see that

$$[s \cdot u] \cdot [t \cdot x] + [t \cdot v] \cdot [s \cdot w] = [t \cdot s] \cdot [u \cdot x] + [t \cdot s] \cdot [v \cdot w]$$
$$= [t \cdot s] \cdot [u \cdot x + v \cdot w]$$
$$= t \cdot (s \cdot [u \cdot x + v \cdot w]),$$

and

$$[t \cdot v] \cdot [t \cdot x] = [t \cdot t] \cdot [v \cdot x] = t \cdot (t \cdot [v \cdot x]),$$

so that the partners of the identifying pair of the second rational are those of the identifying pair of the first rational multiplied by $t \neq 0$. Thus, by an earlier observation,

$$(s \cdot [u \cdot x + v \cdot w], \ t \cdot [v \cdot x]) \curvearrowright ([s \cdot u] \cdot [t \cdot x] + [t \cdot v] \cdot [s \cdot w], \ [t \cdot v] \cdot [t \cdot x]),$$

and

$$R(s \cdot [u \cdot x + v \cdot w], \ t \cdot [v \cdot x]) = R([s \cdot u] \cdot [t \cdot x] + [t \cdot v] \cdot [s \cdot w], \ [t \cdot v] \cdot [t \cdot x]),$$

as we wished to show. Therefore, letting R denote the set of all rationals, we have now shown that the system $\{R; \boxed{+}, \boxed{\times}\}$ is a number system.

In the system $\{R; \boxed{+}, \boxed{\times}\}$ is there any rational which plays the role of the unique multiplicative identity? The answer is yes! The identity rational with respect to multiplication is that rational whose representation has the form $R(u, u)$, where u is any nonzero integer. This means, of course, that any one of the infinitely many symbols

$$\dots, \ R(-2, -2), \ R(-1, -1), \ R(1, 1), \ R(2, 2), \ R(3, 3), \dots$$

is a representation for the unique multiplicative identity. To prove this assertion we need only observe that when $u \neq 0$

$$R(u, u) \ \boxed{\times} \ R(s, t) = R(u \cdot s, u \cdot t) = R(s, t)$$

for any rational $R(s, t)$.

EXERCISE GROUP 8–3

1. Verify the commutative, associative, and distributive laws in the following cases:

(a) $R(5, 7) \ \boxed{+} \ R(-2, 3) = R(-2, 3) \ \boxed{+} \ R(5, 7)$

(b) $R(4, 9) \ \boxed{\times} \ [R(-1, 5) \ \boxed{\times} \ R(2, 4)] = [R(4, 9) \ \boxed{\times} \ R(-1, 5)] \ \boxed{\times} \ R(2, 4)$

(c) $R(4, 3) \ \boxed{\times} \ [R(2, 7) \ \boxed{+} \ R(2, 3)]$
$$= R(4, 3) \ \boxed{\times} \ R(2, 7) \ \boxed{+} \ R(4, 3) \ \boxed{\times} \ R(2, 3)$$

8–4 Subtraction and division. Before discussing the possibility of subtraction and division operations on the rationals we restate some results that have been previously obtained. The rationals have a unique additive identity ("zero" element), and this rational is the one whose representation has the form $R(0, u)$. The rationals have a unique multiplicative identity ("one" element), and this rational is the one whose representation has the form $R(u, u)$. We have also observed that each rational $R(s, t)$ has an *additive inverse* $R(-s, t)$ or $-R(s, t)$, where $-s$ is the integer which is the additive inverse of the integer s.

It seems natural to inquire whether each rational has a multiplicative inverse. Although this would seem desirable, it is not true that each rational has a multiplicative inverse; only the "nonzero" rationals have inverses, where by "nonzero" rational we mean any rational $R(s, t)$ with $s \neq 0$. If $R(s, t)$ is any nonzero rational, then $R(t, s)$ is its multiplicative inverse, for we observe that

$$R(s, t) \; \boxed{\times} \; R(t, s) \; = \; R(s \cdot t, t \cdot s),$$

the multiplicative identity. It is also true that the zero rational, $R(0, u)$, *never* has a multiplicative inverse, for if $R(s, t)$ is *any* rational whatever,

$$R(0, u) \; \boxed{\times} \; R(s, t) \; = \; R(0 \cdot s, u \cdot t) \; = \; R(0, u \cdot t),$$

which is *not* the multiplicative identity. Therefore, it is certainly true that the rationals are *not* a group with respect to multiplication. If this seems disappointing, there is one consolation: *the nonzero rationals are a group with respect to multiplication.*

We define subtraction on the rationals as follows.

DEFINITION. If $R(s, t)$ and $R(u, v)$ are any rationals, then

$$R(s, t) \; - \; R(u, v) \; = \; R(s, t) \; \boxed{+} \; [-R(u, v)],$$

where $-R(u, v) = R(-u, v)$ is the additive inverse of $R(u, v)$.

We call $R(s, t)$ the *subtrahend* and $R(u, v)$ the *minuend*, and our definition says "to subtract rationals, add the additive inverse of the minuend." Note that since *every* rational has an additive inverse, subtraction is *always* possible.

Division on the rationals is now defined.

DEFINITION. Let $R(s, t)$ and $R(u, v)$ be rationals, where $R(u, v)$ is a nonzero rational ($u \neq 0$). Then

$$R(s, t) \; \div \; R(u, v) \; = \; R(s, t) \; \boxed{\times} \; R(v, u) \; = \; R(s \cdot v, t \cdot u).$$

Note that our definition tells us only how to divide by a *nonzero* rational. The rational $R(s, t)$ is called the *dividend*, $R(u, v)$ is called the *divisor*, and $R(s \cdot v, t \cdot u)$ is called the *quotient*. We could have also said that $R(s, t) \div R(u, v)$ is that rational $R(x, y)$, if it exists, such that $R(u, v) \; \boxed{\times} \; R(x, y) = R(s, t)$, and we note that $R(s \cdot v, t \cdot u)$ is that rational, for

$$
\begin{aligned}
R(u, v) \; \boxed{\times} \; R(s \cdot v, t \cdot u) \; &= \; R(u \cdot [s \cdot v], v \cdot [t \cdot u]) \\
&= \; R([u \cdot v] \cdot s, [u \cdot v] \cdot t) \\
&= \; R(s, t).
\end{aligned}
$$

Suppose that we try to divide by a zero rational, $R(s, t) \div R(0, v)$. Let us assume for the moment that this division is possible, so that we can write

$$R(s, t) \div R(0, v) = R(x, y).$$

Then, if $R(x, y)$ is the quotient as we suppose, it must be true that $R(0, v) \boxed{\times} R(x, y) = R(s, t)$. But

$$R(0, v) \boxed{\times} R(x, y) = R(0 \cdot x, v \cdot y) = R(0, v \cdot y),$$

so that $R(0, v \cdot y) = R(s, t)$. This, however, is not true unless $s = 0$ and $R(s, t)$ is a zero rational. Thus it is not possible to divide $R(s, t)$ by a zero rational unless $R(s, t)$ is itself a zero rational. Does this argument imply that it *is* possible to divide a zero rational by a zero rational? The answer is emphatically no! If we write

$$R(0, t) \div R(0, v) = R(x, y),$$

we note that

$$R(0, v) \boxed{\times} R(x, y) = R(0 \cdot x, v \cdot y) = R(0, v \cdot y) = R(0, t),$$

and this holds true regardless of what rational $R(x, y)$ may be. Therefore, although it is technically possible to divide in this case, we see that the quotient is not unique, and we are interested in only those divisions which yield unique quotients.

8–5 The cancellation laws. There are two cancellation laws, one for addition and one for multiplication, which must now be discussed:

(1) If
$$R(s, t) \boxed{+} R(u, v) = R(s, t) \boxed{+} R(w, x),$$
then
$$R(u, v) = R(w, x),$$
and

(2) If $R(s, t)$ is a nonzero rational and
$$R(s, t) \boxed{\times} R(u, v) = R(s, t) \boxed{\times} R(w, x),$$
then
$$R(u, v) = R(w, x).$$

To prove (1) we take

$$R(s, t) \boxed{+} R(u, v) = R(s, t) \boxed{+} R(w, x)$$

and add $R(-s, t)$ to both sides (justify!), obtaining

$$R(-s, t) \boxplus [R(s, t) \boxplus R(u, v)] = R(-s, t) \boxplus [R(s, t) \boxplus R(w, x)],$$
$$[R(-s, t) \boxplus R(s, t)] \boxplus R(u, v) = [R(-s, t) \boxplus R(s, t)] \boxplus R(w, x),$$
$$R(-s \cdot t + t \cdot s, t^2) \boxplus R(u, v) = R(-s \cdot t + t \cdot s, t^2) \boxplus R(w, x),$$
$$R(0, t^2) \boxplus R(u, v) = R(0, t^2) \boxplus R(w, x),$$
$$R(u, v) = R(w, x).$$

(Justify each step.) To prove (2) we multiply both sides of

$$R(s, t) \boxtimes R(u, v) = R(s, t) \boxtimes R(w, x)$$

by $R(t, s)$, the multiplicative inverse of $R(s, t)$, and proceed as we did in proving (1). (Complete the proof.)

It is now clear that the system of rationals with the operations $\boxplus$ and $\boxtimes$, $\{R; \boxplus, \boxtimes\}$, is an integral domain, for $\{R; \boxplus, \boxtimes\}$ is a group with respect to addition, there exist unique and distinct additive and multiplicative identity elements, and the multiplicative cancellation property holds.

8–6 The fractions. We now briefly summarize the steps taken in constructing the rational number system.

(i) We took the set $I \times J$ of all ordered pairs of integers whose right partners are not zero and defined an equivalence relation on $I \times J$, the relation $\curvearrowright$ (wiggle): for any pairs (r, s), $(t, u) \in I \times J$, $(r, s) \curvearrowright (t, u)$ if, and only if, $r \cdot u = s \cdot t$.

(ii) We constructed some equivalence sets. For each pair $(r, s) \in I \times J$ the set $R(r, s)$ was constructed, where $R(r, s)$ contains those pairs, and only those pairs, $(t, u) \in I \times J$ such that $(t, u) \curvearrowright (r, s)$. We called the family R of all such equivalence sets the *rationals*.

(iii) We defined operations $\boxplus$ and $\boxtimes$ on R and proved that $\{R; \boxplus, \boxtimes\}$ is a number system.

Now think for a moment of the sets $I \times J$ and R as being merely *sets of symbols* and let us define the mapping α of $I \times J$ into R as follows:

$$\alpha : (r, s) \rightarrow R(r, s) \quad \text{for each} \quad (r, s) \in I \times J.$$

In other words, α associates each ordered pair (symbol) belonging to $I \times J$ with the symbol belonging to R which has that pair as its identifying pair. We see at once that α is a mapping of $I \times J$ *onto* R, for if we take any symbol $R(t, u) \in R$, there exists a symbol $(t, u) \in I \times J$ such that

$$\alpha : (t, u) \rightarrow R(t, u).$$

Moreover, α is a 1 : 1 mapping of $I \times J$ onto R, for if we take any distinct symbols (r, s), $(t, u) \in I \times J$, these symbols have distinct images $R(r, s)$ and $R(t, u)$. It is true, of course, that the images $R(r, s)$ and $R(t, u)$ might be symbols for the same set, but they are *distinct* symbols.

The fact that there is a 1 : 1 correspondence between the symbols in $I \times J$ and the symbols in R immediately suggests the possibility of using the ordered pairs in $I \times J$ as symbols for the rational numbers in place of the symbols in set R. This is exactly what we will do except that we shall write the pairs (r, s) in $I \times J$ in the form r/s ("r over s"); the set $I \times J$ with all its pairs (r, s) written in the form r/s will be called *fractions*. Of course, if fractions are to be used as symbols for the rational numbers, we must define operations on them, and we *define addition* and *multiplication* in such a way that the 1 : 1 correspondence between the two sets will be an isomorphism with respect to each operation; we shall use the familiar symbols $+$ and $\cdot$ to denote addition and multiplication on the fractions:

$$R(s, t) \; \boxed{+} \; R(u, v) = R(s \cdot v + t \cdot u, t \cdot v)$$
$$\updownarrow \qquad\qquad \updownarrow \qquad\qquad\qquad \updownarrow$$
$$\frac{s}{t} \quad + \quad \frac{u}{v} \quad = \quad \frac{s \cdot v + t \cdot u}{t \cdot v},$$

and

$$R(s, t) \; \boxed{\times} \; R(u, v) = R(s \cdot u, t \cdot v)$$
$$\updownarrow \qquad\qquad \updownarrow \qquad\qquad \updownarrow$$
$$\frac{s}{t} \quad \cdot \quad \frac{u}{v} \quad = \quad \frac{s \cdot u}{t \cdot v}.$$

For *subtraction* and *division* we have

$$R(s, t) \; - \; R(u, v) = R(s, t) \; \boxed{+} \; R(-u, v)$$
$$\updownarrow \qquad\qquad \updownarrow \qquad\qquad \updownarrow \qquad\qquad \updownarrow$$
$$\frac{s}{t} \quad - \quad \frac{u}{v} \quad = \quad \frac{s}{t} \quad + \quad \frac{-u}{v},$$

and

$$R(s, t) \div R(u, v) = R(s, t) \; \boxed{\times} \; R(v, u), \; (u \neq 0)$$
$$\updownarrow \qquad\qquad \updownarrow \qquad\qquad \updownarrow \qquad\qquad \updownarrow$$
$$\frac{s}{t} \quad \div \quad \frac{u}{v} \quad = \quad \frac{s}{t} \quad \cdot \quad \frac{v}{u}.$$

Since our 1 : 1 correspondence is an isomorphism, we know also that the commutative, associative, distributive, and cancellation laws all hold for fractions. The fractions, from now on, will be the set of symbols that we use to represent the rational numbers and we will operate with the fractions according to the foregoing definitions. From now on, when we speak of the rational numbers, we will mean the fractions and we may *redefine* a rational number to be a number of the form p/q, where p and q are integers and $q \neq 0$.

In the fractions, any fraction

$$\cdots, \frac{0}{-3}, \frac{0}{-2}, \frac{0}{-1}, \frac{0}{1}, \frac{0}{2}, \frac{0}{3}, \cdots$$

may be used as the zero element (additive identity) and any fraction

$$\cdots, \frac{-3}{-3}, \frac{-2}{-2}, \frac{-1}{-1}, \frac{1}{1}, \frac{2}{2}, \frac{3}{3}, \cdots$$

may be used as the one element (multiplicative identity). The additive inverse of any fraction u/v is $-u/v$ and the multiplicative inverse of any nonzero fraction u/v is v/u, called the *reciprocal* of u/v. The zero element of the rational numbers (fractions) will ordinarily be written simply 0, it being understood that we may supply any nonzero denominator we wish. With this agreement, we note an obvious $1:1$ correspondence between the integers and a certain subset of the rational numbers:

$$\cdots, -4, -3, -2, -1, 0, 1, 2, 3, 4, \cdots$$
$$\updownarrow \quad \updownarrow \quad \updownarrow \quad \updownarrow \; \updownarrow \; \updownarrow \; \updownarrow \; \updownarrow \; \updownarrow$$
$$\cdots, \frac{-4}{1}, \frac{-3}{1}, \frac{-2}{1}, \frac{-1}{1}, \frac{0}{1}, \frac{1}{1}, \frac{2}{1}, \frac{3}{1}, \frac{4}{1}, \cdots$$

This $1:1$ correspondence is an isomorphism with respect to addition and also with respect to multiplication, for if x and y are any integers such that $x + y = z$ and $x \cdot y = u$, we have

$$\begin{array}{ccc} x + y & = & z \\ \updownarrow \quad \updownarrow & & \updownarrow \\ \dfrac{x}{1} + \dfrac{y}{1} & = & \dfrac{z}{1}, \end{array}$$

and

$$\begin{array}{ccc} x \cdot y & = & u \\ \updownarrow \quad \updownarrow & & \updownarrow \\ \dfrac{x}{1} \cdot \dfrac{y}{1} & = & \dfrac{u}{1}. \end{array}$$

Letting

$$R_1 = \left\{ \cdots, \frac{-4}{1}, \frac{-3}{1}, \frac{-2}{1}, \frac{-1}{1}, \frac{0}{1}, \frac{1}{1}, \frac{2}{1}, \frac{3}{1}, \frac{4}{1}, \cdots \right\},$$

we can conclude that $\{R_1; +, \cdot\}$ is a number system and that the systems $\{I; \oplus, \otimes\}$ and $\{R_1; +, \cdot\}$ have *the same structure*. Because these systems have the same structure any problem requiring only the arithmetic of integers for its solution may be solved using the R_1 arithmetic and vice versa. Thus we ordinarily make no distinction between the integers

$$\cdots, -3, -2, -1, 0, 1, 2, 3, \cdots$$

and the rationals

$$\cdots, \frac{-3}{1}, \frac{-2}{1}, \frac{-1}{1}, \frac{0}{1}, \frac{1}{1}, \frac{2}{1}, \frac{3}{1}, \cdots$$

and we use the symbols

$$\cdots, -3, -2, -1, 0, 1, 2, 3, \cdots$$

for both. It would *appear* as though the set I is a subset of the set R and that we have embedded the integers in the rationals or, putting it another way, that we have extended the integers (review Section 7–7). From now on, then, we shall adopt this point of view.

EXERCISE GROUP 8–4

1. Prove that

$$\frac{-u}{v} = \frac{u}{-v} \quad \text{and} \quad \frac{u}{v} = \frac{-u}{-v}.$$

[*Hint:* This may be done by using the isomorphism $R(s, t) \leftrightarrow (s/t)$ between the rationals and the fractions.]

2. If we are to multiply 6/11 and 44/9, we write

$$\frac{\overset{2}{\cancel{6}}}{\cancel{11}} \cdot \frac{\overset{4}{\cancel{44}}}{\cancel{9}_{3}} = \frac{8}{3}.$$

Justify this "cancellation" procedure by multiplying $R(6, 11)$ and $R(44, 9)$. Also justify reducing 8/12 to 2/3 by writing

$$\frac{\overset{2}{\cancel{8}}}{\cancel{12}_{3}} = \frac{2}{3}.$$

In what sense are 8/12 and 2/3 equal?

8–7 Ordering the rationals. In Section 7–6 we defined the relation $<$ (is less than) on the integers and we called $<$ an order relation. Actually, however, we have *not* proved that the relation $<$ as we defined it is an order relation and we must now do this.

(1) Let x, y, z be any integers such that $x < y$ and $y < z$. Then, it follows by definition of $<$ that $y - x$ and $z - y$ are both positive integers, so that their sum, $(y - x) + (z - y)$, is also a positive integer (why?). Thus we have

$$(y - x) + (z - y) = (z - y) + (y - x) \qquad \text{(Why?)}$$
$$= \{z + (-y)\} + [y + (-x)] \qquad \text{(Why?)}$$
$$= z + \{(-y) + [y + (-x)]\} \qquad \text{(Why?)}$$
$$= z + \{[(-y) + y] + (-x)\} \qquad \text{(Why?)}$$
$$= z + \{0 + (-x)\} \qquad \text{(Why?)}$$
$$= z - x, \qquad \text{(Why?)}$$

so that $z - x$ is a positive integer and $x < z$. We have proved that $<$ is transitive.

(2) For every $x \in I$ it is true that $x - x = 0$; that is, $x - x$ is *not* a positive integer. Thus, $x \not< x$ and $<$ is antireflexive.

(3) If x and y are any integers such that $x < y$, then $y - x$ is a positive integer. Thus $-(y - x)$, the additive inverse of $y - x$, is a *negative* integer. But

$$-(y - x) = (-1) \cdot [y + (-x)]$$
$$= (-1) \cdot y + (-1) \cdot (-x)$$
$$= (-y) + [-(-x)]$$
$$= (-y) + x$$
$$= x + (-y)$$
$$= x - y,$$

so that $y \not< x$ and $<$ is antisymmetric.

We have established that $<$ is an order relation on the integers.

The question now arises: Is it possible to define an order relation on the rationals which is consistent with the relation $<$ defined on the integers? This should not be difficult if we recall our grammar school arithmetic. Imagine that a hungry sixth-grade boy is asked whether he would prefer $\frac{2}{3}$ of a pie or $\frac{5}{8}$ of a pie. If this boy remembers anything about fractions, he will realize at once that he cannot compare $\frac{2}{3}$ and $\frac{5}{8}$ without first expressing them as equivalent fractions having the same denominator, that is, he would compare $\frac{16}{24}$ and $\frac{15}{24}$. There is no need to guess what choice our sixth-grade friend would make and, like him, if we were asked to compare two fractions p/q and r/s, where p, q, r, s are positive integers, we would compare the fractions ps/qs* and qr/qs* by comparing the integers ps and qr.

In this simple case, where all numerators and denominators are positive integers, we might tentatively define a relation $<$ by saying "$p/q < r/s$ if, and only if, $ps < qr$." We could use this definition of $<$ to order the positive rationals, where by "positive rational" we mean any rational number p/q whose numerator and denominator are either both positive or both negative. The rationals $5/7$, $-3/-2$, $12/5$, $-48/-23$ are examples of positive rationals. The definition we finally adopt for the relation $<$ on the rationals must order *all* the rationals, so we consider the following argument.

ARGUMENT. Let p/q and r/s be any rationals. Then, pq/q^2 and rs/s^2 both have positive denominators and are equal, respectively, to p/q and r/s (why?). By writing

$$\frac{pqs^2}{(qs)^2} = \frac{pq}{q^2} \quad \text{and} \quad \frac{rsq^2}{(qs)^2} = \frac{rs}{s^2},$$

we have two rationals which are equal, respectively, to p/q and r/s and which may be compared, since they have the *same* positive denominator.

It would seem reasonable, then, to adopt the following definition for $<$ on the rationals.

DEFINITION. For any rationals p/q and r/s we say that $p/q < r/s$ if, and only if, $pqs^2 < rsq^2$.

We must now verify that the relation $<$, as we have defined it, is an order relation. First of all we *assume* that there is a trichotomy law for integers: If x and y are any integers then one and only one of the statements $x < y$, $x = y$, $x > y$, is true (you should be able to prove this). With this assumption it is immediately clear that $p/q \not< p/q$, so that $<$ is antireflexive, and if

$$\frac{p}{q} < \frac{r}{s} \quad \text{then} \quad \frac{r}{s} \not< \frac{p}{q},$$

so that $<$ is antisymmetric. To show that $<$ is transitive we suppose that p/q, r/s, u/v are rationals such that $p/q < r/s$ and $r/s < u/v$. Thus it is true that

$$pqs^2 < rsq^2 \quad \text{and} \quad rsv^2 < uvs^2,$$

---

* Here we have abandoned the use of the multiplication sign and have written ps and qr instead of $p \cdot s$ and $q \cdot r$. From now on we will use $\cdot$ to indicate multiplication only where necessary for emphasis.

and to prove that $p/q < u/v$, we must show that $pqv^2 < uvq^2$. To prove this we shall need the following proposition: If x, y, z are integers such that $x < y$ and $0 < z$, then $xz < yz$ (Problem 9, Exercise Group 7–5).

Proof. We must show that $yz - xz$ is positive. Using the distributive law, we may write $yz - xz = z(y - x)$. Since it is given that $x < y$ and $0 < z$, it is immediately clear that the factors z and $y - x$ are both positive and their product $z(y - x) = yz - xz$ is positive. This proves the proposition.

Returning now to our proof that $p/q < u/v$, we make use of the two inequalities $pqs^2 < rsq^2$ and $rsv^2 < uvs^2$. Multiplying both sides of the first inequality by the positive integer v^2 and both sides of the second by the positive integer q^2, we obtain $pqs^2v^2 < rsq^2v^2$ and $rsq^2v^2 < uvq^2s^2$. (How do we know that v^2 and q^2 are positive integers?) Therefore $pqs^2v^2 < uvq^2s^2$ (why?). However, this is not exactly what we wanted to prove; we want to show that $pqv^2 < uvq^2$! We do know that one and only one of the statements

$$\text{(i)} \quad pqv^2 < uvq^2,$$
$$\text{(ii)} \quad pqv^2 = uvq^2,$$
$$\text{(iii)} \quad pqv^2 > uvq^2$$

is true. Thus if we can find which two of these three statements are false we will know that the remaining statement is true. The statement

$$\text{(ii)} \quad pqv^2 = uvq^2$$

is false, for if it were true $pqs^2v^2 = uvq^2s^2$ would necessarily have to be a true statement (why?) and we know that it is not. Likewise, we know that the statement

$$\text{(iii)} \quad pqv^2 > uvq^2$$

is false, for if it were true $pqs^2v^2 > uvq^2s^2$ would be a true statement (why?) and we know that it is not. Therefore, of the three statements above, the second and third are false and we conclude that

$$\text{(i)} \quad pqv^2 < uvq^2$$

is true, so that $p/q < u/v$. The transitive property of the relation $<$ on the rationals has been established and $<$ is an order relation on the rationals.

Finally, to show that the order relation $<$ on the rationals is consistent with the order relation $<$ on the integers, we look at the $1:1$ correspondence (isomorphism) between the integers and the set of rationals of the form $p/1$ (page 134). If x and y are any integers such that $x < y$, it is true that $x \cdot 1 \cdot 1^2 < y \cdot 1 \cdot 1^2$, so that $x/1 < y/1$, and, conversely, if $x/1 < y/1$, it is true that $x \cdot 1 \cdot 1^2 < y \cdot 1 \cdot 1^2$, so that $x < y$. Thus, the order relation $<$ on the rationals is consistent with the order relation $<$ on the integers.

A property of the rationals that is closely related to what we have been talking about is the property of *denseness*, which will now be explained. We begin by considering the following proposition: If x and y are any rationals such that $x < y$, then there exists at least one rational z such that $x < z < y$. In more prosaic terms this proposition states: There is at least one rational between any two given rationals. We shall prove this proposition, but first let us observe its implication. Assuming for the moment that the proposition is true, let r_1 and r_2 be any two rationals such that $r_1 < r_2$. There exists a rational r_3 such that $r_1 < r_3 < r_2$; there exist rationals r_4 and r_5 such that $r_1 < r_4 < r_3$ and $r_3 < r_5 < r_2$, so that $r_1 < r_4 < r_3 < r_5 < r_2$, and so on. In short, this process may be carried on indefinitely, and if there exists one rational number between two given rationals, there exist infinitely many!

To prove the proposition we will need these two facts: (i) If x, y, z are integers such that $x < y$, then $x + z < y + z$ (Problem 8, Exercise Group 7–5) and (ii) if x, y, z are integers such that $x < y$ and $0 < z$, then $xz < yz$ (Problem 9, Exercise Group 7–5). All that is needed to prove the above proposition is to exhibit *one* rational number z which lies between given rationals p/q and r/s, where $p/q < r/s$. We assert that the rational number

$$z = \frac{1}{2}\left(\frac{p}{q} + \frac{r}{s}\right) = \frac{ps + qr}{2qs},$$

the arithmetic mean (average) of p/q and r/s, is such a number.

Proof. We must prove two things: (i) that $p/q < (ps + qr)/2qs$ and (ii) $(ps + qr)/2qs < r/s$. To prove (i) we start with the inequality $pqs^2 < rsq^2$, which we *know* to be true, since we are given that $p/q < r/s$. We take the following steps in order:

$$pqs^2 < rsq^2,$$

$$(pqs^2)q^2 < (rsq^2)q^2, \qquad\qquad \text{(Why?)}$$

$$pq^3s^2 < rsq^4, \qquad\qquad \text{(Why?)}$$

$$2pq^3s^2 < 2rsq^4, \qquad\qquad \text{(Why?)}$$

$$4pq^3s^2 < 2pq^3s^2 + 2rsq^4, \qquad\qquad \text{(Why?)}$$

$$pq(2qs)^2 < (ps + qr)(2qs)q^2. \qquad\qquad \text{(Why?)}$$

The last inequality implies that

$$\frac{p}{q} < \frac{ps + qr}{2qs}.$$ (Why?)

To prove (ii) we start with the same inequality and take the following steps in order:

$$pqs^2 < rsq^2,$$

$$(pqs^2)s^2 < (rsq^2)s^2,$$ (Why?)

$$pqs^4 < rs^3q^2,$$ (Why?)

$$2pqs^4 < 2rs^3q^2,$$ (Why?)

$$2pqs^4 + 2rs^3q^2 < 4rs^3q^2,$$ (Why?)

$$(ps + qr)(2qs)s^2 < rs(2qs)^2.$$ (Why?)

The last inequality implies that

$$\frac{ps + qr}{2qs} < \frac{r}{s}.$$ (Why?)

We have proved (i) and (ii) and know that our proposition is true.

Considering this proposition together with its implication, which we discussed earlier, we can say that there are *infinitely many* rational numbers between any two given distinct rational numbers. Mathematicians are in the habit of saying that "the rationals are everywhere dense."

EXERCISE GROUP 8–5

1. Prove the trichotomy law for *integers* as stated on page 137.

2. On page 137 a positive rational was informally defined as a rational whose numerator and denominator have the same sign. Let us now define positive rational by saying that a rational p/q is positive if and only if $0 < p/q$. Show that the two definitions are equivalent.

3. Order the set of rationals

$$\left\{ -3, \frac{13}{6}, -\frac{11}{3}, \frac{2}{5}, \frac{17}{8} \right\}.$$

That is, write the list so that each rational is less than the rational immediately to its right. [*Hint:* Rewrite these rationals so that they have the same positive denominator.]

4. Insert seven rational numbers (in order) between 2/7 and 5/8. Do this also by using common *numerators* instead of common denominators.

8–8 Fields. Mathematicians are very interested in certain mathematical systems which are called *fields*. A field has desirable properties not possessed by either a group or an integral domain. There are various equivalent ways of defining a field and we adopt the following.

DEFINITION. A number system $\{S; \oplus, \odot\}$ is a field if, and only if, it possesses the following properties:

   (i) $\{S; \oplus, \odot\}$ is an integral domain, and

   (ii) the set of nonzero elements of S is a group with respect to $\odot$.

It is not our intention to dwell on the subject of fields but merely to point out that the rational number system is a field (verify this). In Chapter 9 we will study the field of real numbers. As it will turn out, the rational numbers may be thought of as a subset of the real numbers and we will therefore call the rationals a "subfield" of the reals. We mention in passing that the field of real numbers actually has infinitely many subfields and the field of rationals is a subset of every one of them; in this sense the field of rationals is the "smallest" subfield of the reals.

It may seem that the rational number system has every desirable feature we could possibly wish for in a number system. We show in Chapter 9 that this is not true. The rational number system is not complete; in other words, there are "holes" in it. We later will explain more fully exactly what this means and we will indicate how the holes are filled.

CHAPTER 9

THE REAL NUMBERS

9–1 Introduction. An encyclopedia would be required to contain an exhaustive study of the real numbers, but here our development of them and the study of their properties will be exceedingly brief and informal.

We shall need to recall the procedure we followed in constructing both the integers and the rationals. In our study of the integers we first constructed a set which we called the *integers*; to build this set we started with $N \times N$, the set of all ordered pairs of natural numbers, and partitioned $N \times N$ by means of the equivalence relation $\frown$ (curl), calling the elements (subsets) of the partition *integers*. We then proceeded to make a *number system* by properly defining operations on the set of integers. Similarly, in our study of rationals we began by first constructing a set called the *rationals*; to build this set we started with the set $I \times J$, the set of all ordered pairs of integers having nonzero right partners, and partitioned $I \times J$ by means of the equivalence relation $\smile$ (wiggle), calling the elements (subsets) of the partition *rationals*. The *rational number system* was then developed by defining operations on the set of rationals.

The important thing to remember here is that in developing both the integers and the rationals *we built the new by using the old*; that is, we built the integers by using the natural numbers and we built the rationals by using the integers, just as, throughout history, new cities have been built from the rubble of older cities and new civilizations have developed from the ashes of older civilizations. The student of history well knows that each new civilization has borrowed something from its predecessor.

It is our purpose in this chapter to develop a new number system, and in doing so we shall make full use of the rational number system.

9–2 Repeating decimals. To begin, let us remember that any rational is a quotient, or ratio, of two integers, and that by dividing the numerator by the denominator we can convert the rational to decimal form. For example, we have $2 = 2/1 = 2.$, $3/2 = 1.5$, $-27/4 = -6.75$, $13/8 = 1.625$, and $12/5 = 2.4$, all of which are *terminating* decimals. There are other rational numbers, however, which cannot be put in the form of a terminating decimal; some of these are $2/3 = 0.666\ldots$ and $-14/11 = -1.272727\ldots$, which are called "repeating" decimals. When we write $2/3 = 0.6666\ldots$ we have some mental reservations, for we do not know exactly in what sense $2/3$ and $0.6666\ldots$ are equal; all we really know is that when 2 is divided by 3 by the formal division process, the division does not "come out even"—we continue to get a remainder no matter how far the formal process is carried. On the other hand, it is proper to write $13/8 = 1.625$, for 1.625 is merely another way of writing $1 + 6/10 + 2/100 + 5/1000$. Upon adding these, we do obtain the rational number $13/8$.

To dispose of the question that has just been raised, we shall need to recall some facts that may have been forgotten, and so we shall briefly review the meaning of exponents, dealing only with powers of 10. First we recall that $10^1 = 10$, 10^2 ("10 squared" or "10 to the second power") $= 10 \cdot 10 = 100$, 10^3 ("10 cubed" or "10 to the third power") $= 10 \cdot 10 \cdot 10 = 1000$, and so on. In general, if n is a positive integer, 10^n ("10 to the nth power") $= 10 \cdot 10 \cdot 10 \cdot \ldots \cdot 10$ (the product of n 10's); it also should be noted that 10^n is merely a notation that is used for the integer 1 followed by n zeros. Thus, $10^4 = 10,000$, $10^5 = 100,000$, $10^6 = 1,000,000$, and so on. 10^{21} would designate the integer 1 followed by 21 zeros. We call the set of numbers

$$\{10^1, 10^2, 10^3, 10^4, \ldots\}$$

the *positive* integral powers of ten, and we note immediately that this set of numbers is closed under multiplication. That is, ordinary multiplication is an operation on this set, for if m and n are any positive integers, $10^m \cdot 10^n = 10^{m+n}$.

We now ask if this mathematical system has a multiplicative identity or, to put the question differently, does there exist a positive integer k such that $10^k \cdot 10^n = 10^n$ for every positive integer n? Since $10^k \cdot 10^n = 10^{k+n}$, we see that for 10^k to be a multiplicative identity it would be necessary that $k = 0$ and $10^0 = 1$. Therefore we *define* $10^0 = 1$ (review pages 17–19) and annex it to the set

$$\{10^1, 10^2, 10^3, \ldots\},$$

obtaining the set

$$\{10^0, 10^1, 10^2, 10^3, \ldots\},$$

called the *non-negative* integral powers of 10, which has ordinary multiplication as an operation, and which has a multiplicative identity element, namely 10^0.

We now ask if each element in the set of non-negative integral powers of ten has a multiplicative inverse or, in other words, for each element 10^n ($n =$ a non-negative integer) in the set does there exist an element 10^k in the set such that $10^n \cdot 10^k = 10^0$? In answering our own question, we see that if $10^n \cdot 10^k = 10^{n+k} = 10^0$ is to hold true, then $n + k = 0$ must be true and k must necessarily be the *additive inverse* of n, $k = -n$.

But 10^{-n} *has no meaning* as yet and is certainly not an element in the set. Therefore we *define* $10^{-n} = 1/10^n$ for every positive integer n. Thus $10^{-1} = 1/10^1 = 1/10$, $10^{-2} = 1/10^2 = 1/100$, and so on. Moreover, $10^n \cdot 10^{-n} = 10^0$ for every positive integer n. We now annex the elements 10^{-1}, 10^{-2}, 10^{-3}, ... to the set of non-negative integral powers of ten and obtain the set

$$\{\ldots, 10^{-3}, 10^{-2}, 10^{-1}, 10^0, 10^1, 10^2, 10^3, \ldots\},$$

which we call simply the *integral* powers of ten. This set has ordinary multiplication as an operation, has a multiplicative identity element, and each element in the set has a multiplicative inverse, 10^0 being its own inverse. In addition, since the integral powers of ten are merely special rational numbers, multiplication is associative and the set is a group with respect to multiplication.

We are now going to use the integral powers of ten to help us shift decimal points. We know that multiplying a decimal by a positive integral power of ten $(10^1 = 10, 10^2 = 100, 10^3 = 1000, \ldots)$ will shift the decimal *to the right* exactly the number of places specified by the exponent of the power of ten. Similarly, multiplying a decimal by a negative integral power of ten $(10^{-1} = 1/10, 10^{-2} = 1/100, 10^{-3} = 1/1000, \ldots)$ will shift the decimal *to the left* exactly the number of places specified by the exponent.

For example, $10^4 \times 0.0538 = 538.$, and $10^{-3} \times 72.9 = 0.0729$. We use this principle to write decimals in the so-called "scientific notation," a notation that is particularly useful in writing very large or very small decimals. An astronomer, for example, in giving the diameter of a certain galaxy as 240 quadrillion (240,000,000,000,000,000) miles would write 2.4×10^{17} miles, and a physicist would write the diameter of a certain atom as 2.4×10^{-8} cm instead of writing 0.000000024 cm. In short, to write any decimal in scientific notation, we *write the decimal with the decimal point located immediately to the right of the first nonzero digit and multiply by the integral power of ten that would have the effect of shifting the decimal point back to its correct position.* Thus we would write 0.00308 as 3.08×10^{-3} and 310,000 as 3.1×10^5. We may also write repeating decimals in scientific notation. For example, we have

$$0.000282828\ldots = 10^{-4} \times 2.\overline{8},$$

and

$$3749.2673673673\ldots = 10^3 \times 3.7492\overline{673},$$

where the bars written above the 28 and 673 mark the location of the first occurrence of the cycle of digits that repeats indefinitely.

We now hope to convince the reader that there is a $1 : 1$ correspondence between the set of distinct rationals and the set of all repeating decimals. First let us show that every rational p/q corresponds to a unique repeating decimal. If $p = 0$, the rational $p/q = 0$ corresponds to $0.0000\ldots$, *a repeating decimal*. If p/q is a negative rational we would carry out the division by using positive integers and attach a minus sign to our result; thus, with no loss of generality, we may assume that p and q are positive integers. If q divides p evenly we obtain a terminating decimal of the form

$$10^r \times a_0 . a_1 a_2 \ldots a_m,$$

which corresponds to

$$10^r \times a_0 . a_1 a_2 \ldots a_m 0000 \ldots,$$

a repeating decimal. On the other hand, if q does not divide p evenly, the division process must continue indefinitely and the quotient must be a repeating decimal. The reason should be clear: since q does not divide p evenly, no step in the division process will ever yield the remainder 0, but rather will yield a remainder from the set

$$\{1, 2, 3, \ldots, q - 2, q - 1\}.$$

If the division process is continued, step after step, not more than q steps will be required for a remainder which has appeared previously to again appear, and as soon as this happens, a certain cycle of digits in the quotient will start to reappear regularly. Therefore each rational does correspond to a repeating decimal if we agree to make every terminating decimal into a repeating decimal by attaching infinitely many zeros to the right of the last digit.

Now let us show that every repeating decimal corresponds in a formal way to a unique rational number. To do this, let

$$N = 10^r \times a_0.a_1 a_2 \ldots a_m \overline{b_1 b_2 \ldots b_n},$$

where N represents the repeating decimal. We are now going to operate on both sides of this equation. We do not know whether what we are going to do is mathematically legal, but we shall proceed boldly by first multiplying both sides by 10^m. This will give us

$$10^m \times N = 10^r \times a_0 a_1 a_2 \ldots a_m . \overline{b_1 b_2 \ldots b_n}, \qquad \text{(i)}$$

where the decimal point on the right has been shifted and now appears immediately to the *left* of the first cycle. Next we multiply both sides of (i) by 10^n, and obtain

$$10^{m+n} \times N = 10^r \times a_0 a_1 a_2 \ldots a_m \overline{b_1 b_2 \ldots b_n}., \qquad \text{(ii)}$$

where the decimal point on the right has been shifted and now appears immediately to the *right* of the first cycle. Now we note that to the right of the decimal point in both (i) and (ii) nothing will be found but infinitely many repetitions of the cycle $b_1b_2 \ldots b_n$; in other words, the parts of (i) and (ii) lying to the right of their decimal points are exactly the same. We now align the decimal points of (i) and (ii) and subtract as follows:

$$10^{m+n} \times N = 10^r \times a_0a_1a_2 \ldots a_mb_1b_2 \ldots b_n.b_1b_2 \ldots b_n \ldots \qquad \text{(ii)}$$

$$10^m \quad \times N = \qquad\qquad 10^r \times a_0a_1a_2 \ldots a_m.b_1b_2 \ldots b_n \ldots \qquad \text{(i)}$$

$$10^{m+n} \times N - 10^m \times N = 10^r \times a_0a_1a_2 \ldots a_mb_1b_2 \ldots b_n.$$
$$-10^r \times a_0a_1a_2 \ldots a_m. \qquad \text{(iii)}$$

Using the distributive law on both the left and right sides of (iii) yields

$$(10^{m+n} - 10^m) \times N = (a_0a_1a_2 \ldots a_mb_1b_2 \ldots b_n - a_0a_1a_2 \ldots a_m) \times 10^r.$$
$$\text{(iv)}$$

Finally, we divide both sides of (iv) by $(10^{m+n} - 10^m)$ to obtain

$$N = \frac{(a_0a_1 \ldots a_mb_1b_2 \ldots b_n - a_0a_1a_2 \ldots a_m) \times 10^r}{10^{m+n} - 10^m}. \qquad \text{(v)}$$

We have tacitly assumed in this last step that $n \neq 0$, so that $10^{m+n} - 10^m \neq 0$. This is a valid assumption, for taking $n = 0$ would be equivalent to saying that there are 0 digits in the repeating cycle; in other words, that (i), the original decimal, is terminating. If we now look at the right side of (v) we see that both numerator and denominator are integers and N is rational.

We have questioned the legality of the steps taken to obtain (i), (ii), (iii), (iv), and (v). This should not worry us too much, however, for by these steps we have been able to show that each repeating decimal *corresponds* to a unique rational.

It has now been shown that each rational corresponds to a unique repeating decimal and each repeating decimal corresponds to a unique rational; that is, there are "as many" (undefined) rationals as there are repeating decimals. We do not know, however, that the mapping in one direction is the inverse of the mapping in the other direction. To put this another way, if r is a rational which corresponds to a repeating decimal d by the division process, we do *not* know that the sequence of questionable operations applied to d will take us back to r. It is true that the mapping of the repeating decimals onto the rationals is the inverse of the mapping of the rationals onto the repeating decimals. We shall accept this assertion without proof and merely verify it for a single case.

(1) To convert 2/7 to a decimal we apply the division process. Thus we have

$$7 \lfloor 2.0^6 0^4 0^5 0^1 0^3 0^2 0^6 0 \ldots ,$$
$$0.2 \; 8 \; 5 \; 7 \; 1 \; 4 \; 2 \; 8 \ldots$$

so that 2/7 corresponds to $0.\overline{285714}$.

(2) To convert $0.\overline{285714}$ to a rational we let $N = 0.\overline{285714}$, so that

$$10^6 \times N = \overline{285714}. \tag{i}$$

Aligning decimal points and subtracting, we find

$$10^6 \times N = 285714.285714 \ldots$$
$$N = 0.285714 \ldots$$
$$\overline{}$$
$$(10^6 - 1) \times N = 285714. \tag{ii}$$

Dividing both sides of (ii) by $10^6 - 1 = 999{,}999$ finally yields

$$N = \frac{285714}{999999}, \tag{iii}$$

which reduces to 2/7 upon division of the numerator and denominator by 142857.

Our assertion above is not the whole truth, for the reason that every terminating decimal may be written as a repeating decimal in *two* ways, either by affixing infinitely many 0's or by diminishing its last digit by 1 and affixing infinitely many 9's. To illustrate, consider the repeating decimal 9.326000 . . . which, of course, is the rational 9326/1000. We have asserted that 9.326000 . . . and 9.325999 . . . correspond to the same rational. To show this, we let $N = 9.325\overline{9}$ and proceed with our formal process:

$$10^3 \times N = 9325.999 \ldots , \tag{i}$$
$$10^4 \times N = 93259.999 \ldots \tag{ii}$$

Subtracting (i) from (ii) yields

$$10^4 \times N = 93259.999 \ldots$$
$$10^3 \times N = 9325.999 \ldots$$
$$\overline{}$$
$$9000 \times N = 83934 \tag{iii}$$

so that

$$N = \frac{83934}{9000} = \frac{9326}{1000}. \tag{iv}$$

In this special case, then, our remark is true. We will assume it true in general and will remove from the set of all repeating decimals those which, from some point on, have no digits but 9's. Now that this is done there is no possibility of two distinct repeating decimals corresponding to the same rational number and we will have a true 1 : 1 correspondence between the set of all distinct rationals and the repeating decimals.

We now begin to understand in what sense the repeating decimals are to be "numbers"—they are actually to be an alternative set of symbols that may be used in place of the rationals. Earlier we wrote $2/3 = 0.6666\ldots$ with some misgivings, but we now know that "equals" will mean "the same as" for we shall *define* addition and multiplication of repeating decimals in such a way that our 1 : 1 correspondence is an isomorphism with respect to addition and also with respect to multiplication. In short, to add (or multiply) any two repeating decimals we add (or multiply) the two corresponding rationals and take the corresponding decimal as the sum (or product). For example, to add $36.\overline{23}$ and $5.\overline{8}$ we first find the corresponding rationals, which are $3587/99$ and $53/9$, and add these rationals, which gives us

$$\frac{3587}{99} + \frac{53}{9} = \frac{4170}{99}.$$

Upon converting this rational to decimal form we obtain $42.\overline{1}$, which we call the sum and write as

$$36.2323\ldots + 5.8888\ldots = 42.12121\ldots$$

Multiplication, subtraction, and division are to be done by a similar procedure. The isomorphism will preserve the commutative, associative, and distributive laws (Problem 4, Exercise Group 5–3). If we ask whether or not the isomorphism preserves the order relation, the answer will depend on how we define an order relation on the repeating decimals. We define the relation $<$ on the repeating decimals by saying that if d_1 and d_2 are any repeating decimals corresponding to r_1 and r_2, respectively, then $d_1 < d_2$ if, and only if, $r_1 < r_2$. We have defined a true order relation; it is antireflexive, antisymmetric, and transitive (why?). Thus, to determine which of two repeating decimals is less than the other we may simply determine which of their corresponding rationals is less than the other. However, there is a more practical way of comparing the relative sizes of two infinite decimals without first converting them to rational form. This method will now be discussed briefly and without attempt at justification.

First of all, we know that if two repeating decimals are of opposite sign, the negative one is less than the positive one. Next, if one positive rational is less than a second, then the negative (additive inverse) of the second will be less than the negative of the first (why?). Therefore it will be sufficient to discuss the comparison of two *positive* repeating decimals and, furthermore, we can assume that the first nonzero digit of each has the same location relative to the decimal point (why?). Let our repeating decimals be

$$d_1 = 10^r \times a_0.a_1a_2 \ldots a_m\overline{b_1b_2 \ldots b_n},$$

and

$$d_2 = 10^r \times s_0.s_1s_2 \ldots s_h\overline{t_1t_2 \ldots t_k}.$$

Now, $d_1 < d_2$ or $d_1 > d_2$ according as $a_0 < s_0$ or $a_0 > s_0$. If $a_0 = s_0$ we compare a_1 and s_1, in which case $d_1 < d_2$ or $d_1 > d_2$ according as $a_1 < s_1$ or $a_1 > s_1$. If $a_0 = s_0$ and $a_1 = s_1$ we compare a_2 and s_2, in which case $d_1 < d_2$ or $d_1 > d_2$ according as $a_2 < s_2$ or $a_2 > s_2$. This process may be carried as far as necessary, and when we encounter for the first time two unequal digits occupying the same position in relation to the decimal point, we will know which of the two repeating decimals is less than the other. Using this method, we can tell at a glance that

$$5.395843843 \ldots < 5.395858585 \ldots.$$

EXERCISE GROUP 9–1

1. Convert 43/18, −7/12, and 248/11 to repeating-decimal form.
2. Convert −738.2$\overline{53}$ and 0.00294$\overline{11}$ to rational form.
3. Convert 29.5$\overline{32}$ to rational form and then convert the fraction back to decimal form.
4. Determine whether 25/32 is less than or greater than 34/43 by comparing their decimal expansions (corresponding repeating decimals).

9–3 Irrational numbers. The word "irrational," which means "not rational," may lead the reader to believe that we are now about to discover some new kind of numbers. This is far from being the case, for the Greeks discovered the existence of irrationals about 500 B.C., in the sense that they constructed line segments whose lengths could not be measured with a "rational ruler," that is, a ruler whose markings correspond to rational numbers. The Greeks called such line segments *incommensurable*.

By using a compass and ruler together with a given unit of measure, we, too, can construct a line segment whose length cannot be expressed as a rational number of units of measure. To prove this, we shall need to recall the

THEOREM OF PYTHAGORAS: If the two legs and the hypotenuse of a right triangle have lengths of a, b, and c units, respectively, then $a^2 + b^2 = c^2$ (Fig. 9–1).

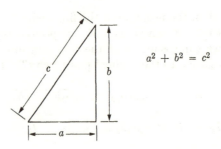

$$a^2 + b^2 = c^2$$

FIGURE 9–1

This theorem is so important that we should prove it and, fortunately, a simple but elegant proof is available to us. To understand this proof we need only know that the area of a square is given by the formula

$$\text{area} = (\text{side})^2,$$

and the area of a triangle is given by the formula

$$\text{area} = \frac{1}{2} \times \text{base} \times \text{altitude}.$$

What we wish to do is to find the length c of the hypotenuse of a right triangle whose legs have lengths a and b. We start by constructing two identical squares having sides of length $a + b$ (Fig. 9–2). The area of square I is divided into five subareas (a square and four right triangles), while the area of square II is divided into six subareas (two squares and four right triangles). It is easily seen that all the right triangles in squares I and II have the same area, namely $\frac{1}{2}ab$. Therefore if we remove the four triangles from both squares, the areas that remain will be equal; in other words,

$$a^2 + b^2 = c^2,$$

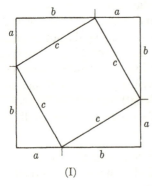

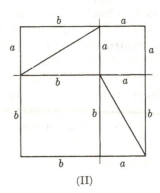

(I)   (II)

FIGURE 9–2

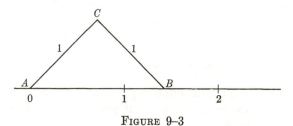

FIGURE 9–3

which we set out to prove. If we now ask what c is, we realize that c is a number whose square is $a^2 + b^2$; that is, c is the so-called positive square root of $a^2 + b^2$, which we designate by $\sqrt{a^2 + b^2}$ and write as $c = \sqrt{a^2 + b^2}$.

We have been assuming all along that a and b are rational numbers, and we ask if $c = \sqrt{a^2 + b^2}$ is always a rational number. The only answer we can give at the moment is that c is *sometimes* rational. If the legs of our right triangle are 3 inches and 4 inches, we write

$$c = \sqrt{3^2 + 4^2} = \sqrt{9 + 16} = \sqrt{25} = 5,$$

and conclude that the number of units (inches) in the length c is a rational number. If we take a right triangle whose legs are both 1 unit in length we write

$$c = \sqrt{1^2 + 1^2} = \sqrt{1 + 1} = \sqrt{2},$$

and conclude that the number of units in the length c is $\sqrt{2}$. What kind of a number is $\sqrt{2}$, or is it a number at all? We can convince ourselves that $\sqrt{2}$ is a number by inspection of Fig. 9–3. Triangle ABC is a right triangle whose legs AC and BC are both 1 unit in length and whose hypotenuse AB is $\sqrt{2}$ units in length. Certainly, unless our eyes deceive us, the hypotenuse AB has a length and, furthermore, this length is more than 1 unit and less than 2 units; no one can ever convince us that this is not true! It's all very well to be so thoroughly convinced that $\sqrt{2}$ is a number, but if it is not a rational number we are going to have to create a new number system which has $\sqrt{2}$ as one of its numbers and on which there is an order relation $<$ that will allow us to write $1 < \sqrt{2} < 2$. To write $1 < \sqrt{2} < 2$ now would be premature.

We are now going to prove that $\sqrt{2}$ is not a rational number, and to do this we shall accept without proof the following theorem.

UNIQUE FACTORIZATION THEOREM. If n is a positive integer, then n may be expressed as the product of prime positive integers and, except for the order in which the integers are multiplied, may be so expressed in only one way.

This theorem says, to put it another way, that for any positive integer n there are prime positive integers $p_1, p_2, \ldots, p_k$ such that

$$n = p_1 \cdot p_2 \cdot \ldots \cdot p_k$$

and, except for multiplying the integers $p_1, p_2, \ldots, p_k$ in a different order, n cannot be expressed as the product of prime positive integers in any other way. For example,

$$2 = 2, \qquad 3 = 3, \qquad 6 = 2 \cdot 3 \quad \text{or} \quad 3 \cdot 2,$$
$$8 = 2 \cdot 2 \cdot 2, \qquad 15 = 5 \cdot 3 \quad \text{or} \quad 3 \cdot 5,$$
$$18 = 3 \cdot 3 \cdot 2 \quad \text{or} \quad 3 \cdot 2 \cdot 3 \quad \text{or} \quad 2 \cdot 3 \cdot 3.$$

Note that when we write $n = p_1 \cdot p_2 \cdot \ldots \cdot p_k$, it is understood that a prime integer might appear more than once in the product. We call the decomposition of n into the product $p_1 \cdot p_2 \cdot \ldots \cdot p_k$ a *factorization* of n, and we call the prime positive integers $p_1, p_2, \ldots, p_k$ *factors* of n. Now, supposing that

$$n = p_1 \cdot p_2 \cdot \ldots \cdot p_k$$

is the unique factorization of n that the theorem guarantees to exist, we obtain, by squaring,

$$n^2 = (p_1 \cdot p_2 \cdot \ldots \cdot p_k) \cdot (p_1 \cdot p_2 \cdot \ldots \cdot p_k)$$
$$= p_1^2 \cdot p_2^2 \cdot \ldots \cdot p_k^2.$$

From this it is immediately apparent that any one of the integers $p_1, p_2, \ldots, p_k$ which occurs as a factor of n must occur as a factor of n^2 an *even number of times*. That is, if p_1, for example, occurs once as a factor of n it will occur twice as a factor of n^2, if p_1 occurs twice as a factor of n it will occur four times as a factor of n^2, and so on. To illustrate, consider the integer 12, which may be written as $12 = 2 \cdot 2 \cdot 3$. The integer 2 occurs *twice* as a prime factor of 12 and the integer 3 *once*, but if we square both sides, we obtain

$$12^2 = (2 \cdot 2 \cdot 3)^2 = 2^2 \cdot 2^2 \cdot 3^2 = 2 \cdot 2 \cdot 2 \cdot 2 \cdot 3 \cdot 3,$$

where the number of occurrences of the factor 2 and the factor 3 have been *doubled*. Now, let us *assume* for the sake of argument that $\sqrt{2}$ is a rational number. Then there are integers p and q such that

$$\frac{p}{q} = \sqrt{2}, \quad \frac{p^2}{q^2} = 2, \quad \text{and} \quad p^2 = 2q^2.$$

We know that if p^2 is expressed as the product of prime integers, then the

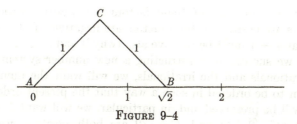

FIGURE 9–4

integer 2 will occur as one of the factors *an even number of times* or *not at all*. Therefore, the integer 2 will appear in the prime factorization of $2q^2$ *an even number of times* or *not at all*. But the integer 2 appears *an even number of times* or *not at all* in the factorization of q^2, and hence appears an *odd* number of times in the factorization of $2q^2$. Having reached this contradiction, we are led to conclude that our assumption is false and that there is no rational number whose square is 2.

We earlier questioned whether $\sqrt{2}$ is actually a number, and now that we have shown there is no rational whose square is 2, the question becomes more urgent than ever. We have defined a number to be an element of a number system and at the present time we know of no number system containing all the rationals and also having $\sqrt{2}$ as one of its elements. Now, in order to explore the possibility of constructing a number system containing both rational and irrational numbers, we shall study the geometric aspects of the problem further by considering a right triangle both of whose legs are 1 unit long.

We suppose that we have a ruler divided into units of the same length as legs AC and BC in Fig. 9–4 and we place the ruler along the hypotenuse as shown. Certainly, we will want the vertex B of the triangle to coincide with a number which tells us the length of AB. Hence we mark $\sqrt{2}$ on our ruler at the point B. Let us now consider the following *proposition*: If r and s are positive rationals, then $r^2 < s^2$ if, and only if, $r < s$.

Proof. To prove this proposition we must show two things: (i) if $r < s$, then $r^2 < s^2$ and (ii) if $r^2 < s^2$, then $r < s$. To prove (i) we assume $r < s$. Then, noting that

$$s^2 - r^2 = (s - r)(s + r),$$

we can argue that $s - r$ is positive since $r < s$, and we can argue that $s + r$ is positive since r and s are given as positive. Thus, since the product of two positive rationals is always a positive rational, $s^2 - r^2$ is positive and $r^2 < s^2$. To prove (ii) we assume $r^2 < s^2$. Again we factor $s^2 - r^2$ and obtain

$$s^2 - r^2 = (s - r)(s + r).$$

Then, since the product of these factors is a positive rational, either both factors are positive or both factors are negative. But $s + r$ is positive, so that $s - r$ must be positive also (why?) and $r < s$.

Now, if we succeed in constructing a new number system containing both the rationals and the irrationals, we will want the numbers in the new system to be ordered in such a way that the present ordering of the rationals will be preserved and, in particular, we will want $1 < \sqrt{2} < 2$ to mean that $\sqrt{2} - 1$ and $2 - \sqrt{2}$ are both positive; we will want $1 < \sqrt{2} < 2$ to mean that $\sqrt{2}$ is between 1 and 2, as in Fig. 9–4; and we will want the proposition we have just proved for the rationals to hold true in the new system of numbers we hope to build. Therefore, in what we do next we shall assume that our real number system has these desirable features.

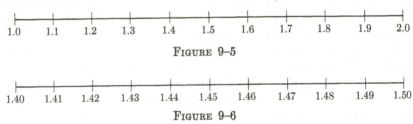

FIGURE 9–5

FIGURE 9–6

We take the ruler we have been using to measure the hypotenuse and we divide the unit length between 1 and 2 into ten equal parts, labeling each division point with the appropriate rational number. Upon magnification, the unit length will appear as shown in Fig. 9–5. We see that these numbers increase as we read them from left to right and that their squares do likewise. Since 1.4 is the largest of these numbers whose square (1.96) is less than $(\sqrt{2})^2 = 2$ and 1.5 is the smallest whose square (2.25) exceeds $(\sqrt{2})^2$, we would like to believe that $1.4 < \sqrt{2} < 1.5$. We now take the section of our ruler between 1.4 and 1.5 and divide it into ten equal parts. After labeling each division point with the appropriate rational number, this section, after magnification, appears as shown in Fig. 9–6. Upon squaring these rational numbers we see that 1.41 is the largest whose square is less than 2 and 1.42 is the smallest whose square is greater than two. We would like to believe that $1.41 < \sqrt{2} < 1.42$.

One might now wonder if this subdividing process, if carried far enough, will ever come to an end. A little reflection tells us that it will not. We have already shown that $\sqrt{2}$ is *not* rational and if the process we began above did terminate, it would necessarily have to terminate with $\sqrt{2}$ equal to a rational number. We know any prolonged continuation of this process in an attempt to find $\sqrt{2}$ is doomed to failure. We would now guess that $\sqrt{2}$ is a number which occupies one of the "holes" we mentioned earlier (Section 8–8) and that there is no hope of filling this hole with a rational number. However, since each rational number is equal to a repeating decimal, the possibility that each irrational is somehow "equal" to an infinite *nonrepeating* decimal suggests itself. As we start exploring this possibility we visualize a number system whose elements are all the infinite decimals, repeating and nonrepeating. We can even visualize $\sqrt{2}$ as equal to an infinite nonrepeating decimal that starts out 1.41 ..., as our inequality $1.41 < \sqrt{2} < 1.42$ suggests. But difficulties arise which at present we are unable to overcome. For example, how can we define addition and multiplication operations on the set of infinite decimals that will be consistent with our present addition and multiplication operations on the rationals, and how can we be sure that the commutative, associative, and distributive laws which apply to the rationals will be preserved? We attempt to answer these questions in the next section.

EXERCISE GROUP 9–2

1. Prove that $\sqrt{3}$, $\sqrt{5}$, and $\sqrt{6}$ are irrational by using the method that was used in proving $\sqrt{2}$ irrational. Using the unique factorization theorem, prove that $\sqrt[3]{4}$ is irrational.

2. By dividing the section of our ruler between 1 and 2 into 10 equal parts and then dividing the section between 1.4 and 1.5 into 10 equal parts we inferred that $1.41 < \sqrt{2} < 1.42$. Carry out this subdivision process two more steps and obtain a similar inference.

9–4 Sequences of rationals. In this section and in the two which follow we will be working with certain kinds of sets of rational numbers which we call *sequences*. We now define what we mean when we speak of a sequence of rational numbers (review Sections 4–1 and 4–2).

DEFINITION. Let P be the set of all positive integers, let R be the set of all rationals, and let α be a mapping of P into R. Then, if the membership list of the set of images (the *range* of α) is written in the order

$$\alpha 1, \alpha 2, \alpha 3, \alpha 4, \alpha 5, \ldots ,$$

we call the membership list a "sequence" of rationals.

In other words, the set of images will be called a *sequence* if the image of 1 is written *first* in the membership list, the image of 2 is written *second*, the image of 3 is written *third*, and so on. Many sequences that we encounter will have the first few terms listed and it will be apparent for these sequences that there is some discernible pattern or rule that can be used to list as many subsequent terms as we wish. The sequence

$$1, \frac{1}{2}, \frac{1}{3}, \frac{1}{4}, \frac{1}{5}, \cdots$$

is an example. The rule appears to be "the nth term is $1/n$." Another sequence of this kind is

$$\frac{1}{2}, \frac{3}{4}, \frac{5}{6}, \frac{7}{8}, \frac{9}{10}, \ldots,$$

where the rule for writing any term appears to be "the nth term is $(2n - 1)/2n$." It is common in mathematical writing to use symbols such as a_n, b_n, c_n, etc. to denote the nth or general terms of sequences. If we let $a_1, a_2, a_3, \ldots$ represent the terms of the sequence $1, 1/2, 1/3, \ldots$ we may state the rule which tells us what term occupies a particular position by writing $a_n = 1/n$. If the symbols $b_1, b_2, b_3, \ldots$ are used to represent the terms of the sequence

$$\frac{1}{2}, \frac{3}{4}, \frac{5}{6}, \ldots,$$

we may write $b_n = (2n - 1)/2n$. Using this rule, we may find

$$b_{12} = \frac{2 \cdot 12 - 1}{2 \cdot 12} = \frac{23}{24}, \qquad b_{21} = \frac{2 \cdot 21 - 1}{2 \cdot 21} = \frac{41}{42},$$

or any term we wish to know.

Another sequence, a trivial one, is $2, 2, 2, 2, \ldots$ and if the symbols $c_1, c_2, c_3, \ldots$ are used to represent the terms of this sequence we may write $c_n = 2$ as our rule for the general term. It is important to understand that for a given sequence, say $d_1, d_2, d_3, d_4, \ldots$, there does *not* have to be a discernible rule that reveals itself to us by a mere glance at the first few terms and that we may use to find the term occupying any specified position. Imagine that we have a wheel of chance with the numbers $0, 1, 2, \ldots, 8, 9$ equally spaced around the circumference. We can use this wheel to construct a sequence. Suppose the wheel stops at 5 on the first spin and we take 5 as the *first* term of our sequence. Suppose the wheel stops at 7 on the next spin and we take 5.7 as the *second* term of our sequence. Suppose the wheel stops at 2 on the next spin and we take 5.72 as the *third* term of our sequence. We can *imagine* spinning the wheel infinitely many times and thereby generating a sequence of rational numbers which starts out $5, 5.7, 5.72, \ldots$ Certainly, after determining a particular term of this sequence we cannot know what the next term will be until we have actually spun the wheel another time. But just as certainly, each term of this sequence does occupy one specified position in the listing. To have a sequence, then, all we need to have is an infinite list of rational numbers with each number occupying a specified position in the list; there may or may not be a discernible rule for telling what rational number occupies any particular position and the terms of the sequence may or may not be distinct.

Before proceeding with further discussion of sequences we shall need to study the concept of the *absolute value* of a rational number.

DEFINITION.   If r is any rational number we denote *the absolute value of r* by the symbol $|r|$, which we define by

$$|r| = \max \{r, -r\}.$$

We know that every rational number r has a unique negative (additive inverse) which we denote by $-r$. Thus our definition tells us: "To find $|r|$, take the number r and its additive inverse and select the larger of the two." For example,

$$\left|\frac{5}{12}\right| = \max \left\{\frac{5}{12}, -\frac{5}{12}\right\} = \frac{5}{12},$$

$$|3| = \max \{3, -3\} = 3,$$

$$|-2| = \max \{-2, -(-2)\} = 2,$$

$$\left|-\frac{2}{3}\right| = \max \left\{-\frac{2}{3}, -\left(-\frac{2}{3}\right)\right\} = \frac{2}{3},$$

$$|0| = \max \{0, -0\} = 0.$$

We have written

$$|-2| = \max \{-2, -(-2)\} = 2$$

and

$$\left|-\frac{2}{3}\right| = \max \left\{-\frac{2}{3}, -\left(-\frac{2}{3}\right)\right\} = \frac{2}{3},$$

which implies that $-(-2) = 2$ and $-(-2/3) = 2/3$. If there are lingering doubts about this, the proof of the following simple proposition should dispel them: If r is any rational number, then $-(-r) = r$.

Proof. The set R of rationals is a group with respect to addition, so that if r is a rational number, r has a unique additive inverse, $-r$, which is also a rational number having a unique additive inverse, $-(-r)$. Therefore the equation $r + (-r) = 0$ tells us not only that $-r$ is the inverse of r, but also that r is the inverse of $-r$; that is, $r = -(-r)$. Therefore it is true that $-(-2) = 2$ and $-(-2/3) = 2/3$, and since max $\{-2, 2\} = 2$ and max $\{-2/3, 2/3\} = 2/3$, we have $|-2| = 2$ and $|-2/3| = 2/3$.

The statement $|0| = \max \{0, -0\} = 0$ possibly might not be clear. However, note that -0 is merely a symbol for the additive inverse of 0, but since 0 is its own additive inverse and the additive inverse of every rational number is unique, $-0 = 0$. Thus max $\{0, -0\}$ means to take

the larger of two equal rationals. The reader may argue that *neither* is
the larger, but he can just as logically argue that each is the larger, since
neither is the smaller. $|0| = 0$ may be taken as a *definition* of $|0|$, and this
definition should seem reasonable if we consider the concept of absolute
value from a geometric standpoint.

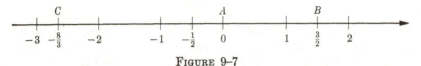

FIGURE 9–7

Imagine that we have a ruler with the rational numbers marked on it,
the positive numbers to the right of 0 and the negative numbers to the
left of 0 (Fig. 9–7). Any rational number r corresponds to a point on the
ruler which is r units from 0, to the right of 0 if r is a positive rational, to
the left of 0 if r is a negative rational, and at 0 if $r = 0$. In order not to
clutter our ruler only a few rational numbers have been marked on it.
The arrow indicates which direction is positive. Now, the distance *from*
A to B is 3/2 units while the distance *from A to C* is −8/3 units. When
we say that the distance *from A to C* is −8/3, obviously we do not mean
that this distance is less than zero but that it has been measured in a
negative direction. If, on the other hand, we ask for the distance *between*
the points A and C, we are not specifying in what direction we are to
measure; the distance *between* these two points is simply 8/3. Thus, if r
is any rational number marked on the ruler we will say that the distance
from 0 to r is r, but that the distance *between* 0 and r is $|r|$; that is, we in-
terpret $|r|$ to mean simply the length of a segment of ruler with 0 at one
end of the segment and r at the other. In particular, $|0|$ may be interpreted
as the length of a segment of our ruler having 0 at *both* ends; in other words,
a segment that ends where it starts—a segment of zero length.

We can prove that absolute values of rational numbers have the following
interesting and useful properties:

$$
\begin{array}{ll}
\text{(i)} & |r| \geq 0, \\
\text{(ii)} & |r| \geq r, \\
\text{(iii)} & |r \cdot s| = |r| \cdot |s|, \\
\text{(iv)} & |r/s| = |r|/|s|, \\
\text{(v)} & |r|^2 = |r^2|, \\
\text{(vi)} & |r + s| \leq |r| + |s| \quad \text{(the ``triangle inequality'')}, \\
\text{(vii)} & |r - s| \leq |r| + |s|.
\end{array}
$$

The proof of (i) is immediate: if $r \neq 0$, then, by definition, $|r|$ is the
positive (larger) member of the set $\{r, -r\}$, whereas if $r = 0$, $|r| = 0$.
Therefore, in any case $|r| \geq 0$ ($\geq$ means "is greater than or equal to").

To prove (ii) we simply note that $|r| = \max \{r, -r\}$. If r is the larger of r and $-r$, then $|r| = r$, whereas if r is the smaller of r and $-r$, then $|r| > r$. Also, if $r = 0$, then $|r| = r$. In any case, $|r| \geq r$.

Property (iii) may be proved by considering the different cases that may occur: r and s both positive, r and s both negative, r positive and s negative and vice versa, and one or both of the numbers r and s equal to zero. Property (iv) may also be proved by considering all possible cases. It is left as an exercise to prove (iii) and (iv).

Property (v) follows at once from (iii). Note that

$$|r|^2 = |r| \cdot |r| \quad \text{and} \quad |r^2| = |r \cdot r|.$$

Since $|r \cdot s| = |r| \cdot |s|$ holds true for *any* rationals r and s, it certainly holds true if $s = r$, in which case we have

$$|r \cdot r| = |r| \cdot |r| \quad \text{or} \quad |r^2| = |r|^2.$$

Property (vi) will be most important to us in later work and a thorough understanding of it is necessary. The inequality

$$|r + s| \leq |r| + |s|$$

($\leq$ means "is less than or equal to") is commonly called the triangle inequality, because in a certain other number system (the complex number system), $|r + s|$, $|r|$, and $|s|$ may be interpreted as the lengths of the sides of a triangle. We will call $|r + s| \leq |r| + |s|$ the triangle inequality from now on.

Proof.
$$|r + s|^2 = |(r + s)^2| = (r + s)^2, \qquad \text{(Why?)}$$
$$(r + s)^2 = r^2 + 2rs + s^2 \leq r^2 + |2rs| + s^2, \qquad \text{(Why?)}$$
$$r^2 + |2rs| + s^2 = |r^2| + 2|r|\,|s| + |s^2|, \qquad \text{(Why?)}$$
$$|r^2| + 2|r|\,|s| + |s^2| = |r|^2 + 2|r|\,|s| + |s|^2, \qquad \text{(Why?)}$$
$$|r|^2 + 2|r|\,|s| + |s|^2 = (|r| + |s|)^2. \qquad \text{(Why?)}$$

Then, since the relations $=$ and $<$ are both transitive, we may write

$$|r + s|^2 \leq (|r| + |s|)^2,$$

so that finally, by the proposition proved on page 153, we conclude that $|r + s| \leq |r| + |s|$.

Property (vii) is an immediate consequence of the triangle inequality. Since $|r + s| \leq |r| + |s|$ is true for *any* rationals r and s, it is certainly true for rationals r and $-s$; that is, $|r + (-s)| \leq |r| + |-s|$. Now $|-s| = \max \{-s, -(-s)\} = \max \{-s, s\} = |s|$ and $r + (-s) = r - s$, so that, upon substituting, we find $|r - s| \leq |r| + |s|$. A more complete understanding of (vi) and (vii) may be gained by visualizing r and s as points on the rational ruler (Fig. 9–7) in different relative positions.

With some understanding of the absolute value concept we are ready to go ahead with our discussion of sequences. The first type of sequence that we need to study is the so-called *null sequence*, which we now define.

DEFINITION. A sequence a_1, a_2, a_3, a_4, ... is called a *null* sequence if, and only if, for any positive integer k a term, a_T, can be found such that $|a_T|$, $|a_{T+1}|$, $|a_{T+2}|$, ... are all less than 10^{-k}.

Note that we are defining a null sequence in terms of the sequence 10^{-1}, 10^{-2}, 10^{-3}, 10^{-4}, ..., which itself is a null sequence. In discussing the sequence

$$10^{-1} = 0.1,\ 10^{-2} = 0.01,\ 10^{-3} = 0.001, \ldots$$

we do not need absolute value signs, since all terms are positive. If we take any positive integer k, a corresponding term 10^{-k} is determined. Since the *very next term*, 10^{-k-1}, and *all terms that follow it* are less than 10^{-k}, and since this will be true for *any* positive integer k that we may select, 10^{-1}, 10^{-2}, 10^{-3}, 10^{-4}, ... is, by definition, a null sequence. As an example, let us consider the sequence

$$-1, \frac{1}{2}, -\frac{1}{3}, \frac{1}{4}, -\frac{1}{5}, \ldots,$$

whose general term is given by $a_n = (-1)^n/n$. By assigning the values $1, 2, 3, 4, \ldots$ to n, in succession, we obtain

$$a_1 = \frac{(-1)^1}{1} = -1, \qquad a_2 = \frac{(-1)^2}{2} = \frac{1}{2},$$

$$a_3 = \frac{(-1)^3}{3} = -\frac{1}{3}, \qquad a_4 = \frac{(-1)^4}{4} = \frac{1}{4}, \ldots$$

As we scan the terms of this sequence, from left to right, we observe that the terms get progressively smaller in absolute value; we *suspect* that this sequence is a null sequence. To confirm our suspicion we take *any* positive integer k and ask ourselves whether or not we can make $|(-1)^n/n| < 10^{-k}$ by taking n sufficiently large; that is, by going out far enough in the sequence. We note that

$$\left| \frac{(-1)^n}{n} \right| = \frac{|(-1)^n|}{|n|}. \qquad \text{(Why?)}$$

Also, since $(-1)^n$ is -1 or 1 according as n is odd or even, and since n is positive, $|(-1)^n| = 1$ and $|n| = n$. Therefore $|(-1)^n/n| = 1/n$ and determining whether our sequence is a null sequence reduces to determining whether we can make $1/n < 10^{-k}$. It is immediately seen that this inequality is true if we take $n = 10^k + 1$ or larger for

$$\frac{1}{10^k + 1} < \frac{1}{10^k} = 10^{-k}. \qquad \text{(Why?)}$$

Thus for *any* positive integer k a term can be found such that it and all following terms have absolute value less than 10^{-k}, and the sequence in question is a null sequence. The following are other examples of null sequences:

$$0, 0, 0, 0, 0, \ldots \qquad (a_n = 0),$$

$$1, \frac{1}{4}, \frac{1}{9}, \frac{1}{16}, \ldots \qquad \left(a_n = \frac{1}{n^2}\right),$$

$$\frac{1}{2}, \frac{1}{4}, \frac{1}{8}, \frac{1}{16}, \ldots \qquad \left(a_n = \frac{1}{2^n}\right).$$

A second type of sequence that will frequently concern us is called the "C-sequence" in honor of Augustin Louis Cauchy, a famous 19th century French mathematician.

> DEFINITION. A sequence $a_1, a_2, a_3, a_4, \ldots$ is called a *C-sequence* if, and only if, for any positive integer k there exists a corresponding positive integer T such that
> $$|a_n - a_m| < 10^{-k}$$
> whenever $m, n > T$.

In less formal language this definition states that $a_1, a_2, a_3, a_4, \ldots$ is a C-sequence if, and only if, for any preselected positive rational 10^{-k}, however small, a term a_T can be found such that any two terms beyond (to the right of) a_T have a difference whose absolute value is less than 10^{-k}. In short, we can always find a point in a C-sequence beyond which any two terms have a difference whose absolute value is as small as we please. An immediate trivial example of a C-sequence is

$$5, 5, 5, 5, \ldots \qquad (a_n = 5).$$

We see that $a_1, a_2, a_3, a_4, \ldots$ are all equal to 5, and that for any positive integers m and n it is true that

$$a_n - a_m = 0 < 10^{-k}$$

for any positive integer k we may select. In fact, we can take any rational number r and construct the sequence $r, r, r, r, \ldots$, which will always be a C-sequence, although a trivial one to be sure.

Before talking about C-sequences that are not trivial we shall find it convenient to introduce a different but equivalent way of characterizing null sequences. If we wish to state that a certain sequence $a_1, a_2, a_3, \ldots$ is a null sequence we will write $a_n \to 0$ as $n \to \infty$, to be read "a_n *approaches* zero as n *becomes* infinite." One should not take this statement too literally; $a_n \to 0$ as $n \to \infty$ means nothing more nor less than "a_n can be made to get and remain as near zero *as we wish* by taking n sufficiently large" or "for *any* positive integer k we can make $|a_n| < 10^{-k}$ for all sufficiently large values of the positive integer n." Often, in the interest of brevity, we will simply write $a_n \to 0$ when we wish to state that a sequence a_1, a_2, a_3, ... is a null sequence. Thus, for the sequence $-1, 1/2, -1/3, 1/4, \ldots$, whose general term is $a_n = (-1)^n/n$, saying $a_n \to 0$ will be our way of saying that this is a null sequence. Therefore, we may use similar language and write "$(b_n - b_m) \to 0$ as $m, n \to \infty$" or simply "$(b_n - b_m) \to 0$" with the phrase "as $m, n \to \infty$" ("as m and n become infinite") omitted for the sake of brevity.

Let us consider the sequence

$$\frac{3}{2}, \frac{5}{4}, \frac{9}{8}, \frac{17}{16}, \ldots,$$

where $a_n = 1 + 1/2^n$. Is this a C-sequence? If we are to show that it is, we must show that $(a_n - a_{m+n}) \to 0$ as $m, n \to \infty$. First we observe that

$$|a_n - a_m| = \left| \left(1 + \frac{1}{2^n}\right) - \left(1 + \frac{1}{2^m}\right) \right| = \left| \frac{1}{2^n} - \frac{1}{2^m} \right|.$$

Since, by (i) and (vii), page 158,

$$\left| \frac{1}{2^n} - \frac{1}{2^m} \right| \geq 0 \quad \text{and} \quad \left| \frac{1}{2^n} - \frac{1}{2^m} \right| \leq \frac{1}{2^n} + \frac{1}{2^m},$$

we may write

$$0 \leq \left| \frac{1}{2^n} - \frac{1}{2^m} \right| \leq \frac{1}{2^n} + \frac{1}{2^m}.$$

Now, $1/2^n \to 0$ as $n \to \infty$ and $1/2^m \to 0$ as $m \to \infty$ (why?) so that $|a_n - a_m|$ is "squeezed" between 0 and the sum of two positive rationals which can be made as small as we please by taking m and n sufficiently large. Thus we can say that $(a_n - a_m) \to 0$ as $m, n \to \infty$ and conclude that the sequence is a C-sequence.

By this time the reader may suspect that we are going to use sequences in some way to build a new number system; this indeed is the case. At this point, then, it will be helpful to have a more convenient notation for sequences. We will simply use one of the letters u, v, w, x, y, z to denote a sequence and we will use the same letter with subscripts attached to represent the terms of the sequence. Thus if x denotes a sequence, x_1, $x_2, x_3, x_4, \ldots$ will be the terms of the sequence.

Since we can add and subtract rational numbers, we can easily define addition and subtraction of sequences by means of our new notation.

DEFINITION. If x is the sequence $x_1, x_2, x_3, x_4, \ldots$ and y is the sequence $y_1, y_2, y_3, y_4, \ldots$ we define $x + y$ to be the sequence

$$x_1 + y_1, \ x_2 + y_2, \ x_3 + y_3, \ldots$$

and $x - y$ to be the sequence

$$x_1 - y_1, \ x_2 - y_2, \ x_3 - y_3, \ldots$$

Thus, by definition, the sum (or difference) of any two given sequences of rational numbers is a sequence of rational numbers each of whose terms is the sum (or difference) of the corresponding terms of the given sequences. Of particular interest to us here is the fact that the sum or difference of any two C-sequences is always a C-sequence; in other words, *the set of all C-sequences is closed under addition and subtraction.*

Proof. Let x and y be C-sequences

$$x_1, x_2, x_3, \ldots \quad \text{and} \quad y_1, y_2, y_3, \ldots,$$

so that

$$x_1 + y_1, \ x_2 + y_2, \ x_3 + y_3, \ldots$$

is their sum, $x + y$, and

$$x_1 - y_1, \ x_2 - y_2, \ x_3 - y_3, \ldots$$

is their difference, $x - y$. To prove that $x + y$ is a C-sequence we consider

$$|(x_n + y_n) - (x_m + y_m)|,$$

which may be written

$$|(x_n - x_m) + (y_n - y_m)|. \quad \text{(Why?)}$$

By (i) and (vi), page 158, we immediately have

$$0 \le |(x_n - x_m) + (y_n - y_m)| \le |x_n - x_m| + |y_n - y_m|.$$

Since x and y are C-sequences, $(x_n - x_m) \to 0$ and $(y_n - y_m) \to 0$ as $m, n \to \infty$. Thus

$$|(x_n + y_n) - (x_m + y_m)|$$

may be "squeezed" between 0 and the sum of two non-negative rationals which may be made as small as we please by taking m and n sufficiently large. Therefore $x + y$ is clearly a C-sequence.

To prove that $x - y$ is a C-sequence we consider

$$|(x_n - y_n) - (x_m - y_m)|,$$

which may be written

$$|(x_n - x_m) - (y_n - y_m)|. \quad \text{(Why?)}$$

By (i) and (vii), page 158, we immediately have

$$0 \le |(x_n - x_m) - (y_n - y_m)| \le |x_n - x_m| + |y_n - y_m|.$$

By an argument similar to that just used it is then apparent that $x - y$ is a C-sequence.

We define the product of two sequences as follows.

DEFINITION. If x and y are the sequences $x_1, x_2, x_3, \ldots$ and $y_1, y_2, y_3, \ldots$, we call the sequence

$$x_1 \cdot y_1, x_2 \cdot y_2, x_3 \cdot y_3, \ldots$$

their product and denote it by $x \cdot y$.

We would now like to prove that the product of two C-sequences is always a C-sequence but we cannot do this without first defining a *bounded* sequence and showing that every C-sequence is bounded.

DEFINITION. A sequence of rationals $x_1, x_2, x_3, \ldots$ is said to be *bounded* if, and only if, there exists a positive rational B such that $|x_n| \le B$ for all values of the positive integer n.

The number B is called a *bound* and, of course, if B is a bound, then $B + 2/3, B + 5, B + 100, \ldots$ are also bounds. In short, if B is a bound then the rational obtained by adding any positive rational to B is also a bound. Some sequences we have looked at are bounded, such as the sequence $3/2, 5/4, 9/8, 17/16, \ldots$, whose general term is $1 + 1/2^n$. For this sequence $3/2$ is a bound but so is $2, 5.793, 8/3, \ldots$ or any rational greater than $3/2$. Thus, if we should wish to show that a sequence of rationals $x_1, x_2, x_3, \ldots$ is bounded, it will be sufficient for us to exhibit *one* positive rational B such that $|x_n| \le B$ for all values of the positive integer n. We now prove the following proposition: *Every C-sequence is bounded.*

Proof. Let x be any C-sequence and select any positive integer k. Corresponding to the selected positive integer k there is a positive integer T such that

$$|x_n - x_m| < 10^{-k}$$

whenever $m, n > T$. Let us now observe that

$$|x_n| = |(x_n - x_m) + x_m| \leq |x_n - x_m| + |x_m|$$

holds true for all positive integers m and n and, in particular, if $m = T + 1$, we have

$$|x_n| = |(x_n - x_{T+1}) + x_{T+1}| \leq |x_n - x_{T+1}| + |x_{T+1}|.$$

Since

$$|x_n - x_{T+1}| < 10^{-k}$$

when $n > T$, we have

$$|x_n| < 10^{-k} + |x_{T+1}|$$

whenever $n > T$. In other words, the terms

$$|x_{T+1}|, |x_{T+2}|, |x_{T+3}|, |x_{T+4}|, \ldots$$

are all less than $10^{-k} + |x_{T+1}|$; *the sequence consisting of all terms beyond x_T is bounded.* Finally, then, if we select a positive rational B which is equal to or greater than the largest positive rational in the finite set

$$\{|x_1|, |x_2|, \ldots, |x_T|, 10^{-k} + |x_{T+1}|\},$$

it will be true that $|x_n| \leq B$ for *every* positive integer n. Our C-sequence is bounded, as we wished to show.

We are now in position to prove that the product, $x \cdot y$, of any two C-sequences is always a C-sequence.

Proof. Let x and y be any C-sequences. To show that $x \cdot y$ is a C-sequence, we must show that

$$|x_n \cdot y_n - x_m \cdot y_m| \to 0 \quad \text{as} \quad m, n \to \infty,$$

and in order to do this we first observe that

$$|x_n \cdot y_n - x_m \cdot y_m| = |(x_n \cdot y_n - x_n \cdot y_m)$$
$$+ (x_n \cdot y_m - x_m \cdot y_m)| \qquad \text{(Why?)}$$
$$= |x_n(y_n - y_m) + y_m(x_n - x_m)| \qquad \text{(Why?)}$$
$$\leq |x_n||y_n - y_m| + |y_m||x_n - x_m|. \qquad \text{(Why?)}$$

Since x and y are C-sequences and are bounded, there exist positive rationals A and B such that $|x_n| \leq A$ for all n, $|y_m| \leq B$ for all m, and

$$|x_n||y_n - y_m| + |y_m||y_n - y_m| \leq A|y_n - y_m| + B|x_n - x_m|.$$

Finally, then, we may write

$$0 \leq |x_n \cdot y_n - x_m \cdot y_m| \leq A|y_n - y_m| + B|x_n - x_m|. \quad \text{(Why?)}$$

Therefore, since

$$(x_n - x_m) \to 0 \qquad \text{and} \qquad (y_n - y_m) \to 0$$

as $m, n \to \infty$, $|x_n \cdot y_n - x_m \cdot y_m|$ may be squeezed between 0 and the sum of two non-negative rationals which may be made as small as we please by taking m and n sufficiently large. In other words,

$$(x_n \cdot y_n - x_m \cdot y_m) \to 0 \quad \text{as} \quad m, n \to \infty$$

and $x \cdot y$ is a C-sequence; the set of all C-sequences is closed under multiplication.

We will omit any discussion of division of sequences at this point, since no useful purpose would be served.

9–5 The real numbers. The C-sequences of rational numbers will be the building blocks that we use in constructing a set to be called the *reals*, and after defining operations on the reals, we will have the *real number system*—a number system without holes. We let

$$C = \{u, v, w, x, y, \ldots\}$$

be the set of all C-sequences and define the following relation $\Re$ on C.

DEFINITION. For any elements $x, y \in C$, $x\Re y$ if, and only if, $(x_n - y_n) \to 0$ as $n \to \infty$.

In other words, one C-sequence has the relation $\Re$ to another C-sequence if their difference is a null sequence. As an example of two such sequences, consider

$$x: \quad \frac{3}{2}, \frac{7}{4}, \frac{15}{8}, \ldots \quad \text{or} \quad 2 - \frac{1}{2}, 2 - \frac{1}{4}, 2 - \frac{1}{8}, \ldots$$

and

$$y: \quad \frac{5}{2}, \frac{9}{4}, \frac{17}{8}, \ldots \quad \text{or} \quad 2 + \frac{1}{2}, 2 + \frac{1}{4}, 2 + \frac{1}{8}, \ldots$$

whose general terms are $x_n = 2 - 1/2^n$ and $y_n = 2 + 1/2^n$. We assert that x and y are both C-sequences, for

$$0 \le |x_n - x_m| = \left|\left(2 - \frac{1}{2^n}\right) - \left(2 - \frac{1}{2^m}\right)\right|$$

$$= \left|-\frac{1}{2^n} + \frac{1}{2^m}\right| \le \frac{1}{2^n} + \frac{1}{2^m},$$

and

$$0 \le |y_n - y_m| = \left|\left(2 + \frac{1}{2^n}\right) - \left(2 + \frac{1}{2^m}\right)\right|$$

$$= \left|\frac{1}{2^n} - \frac{1}{2^m}\right| \le \frac{1}{2^n} + \frac{1}{2^m}.$$

Since $1/2^n \to 0$ as $n \to \infty$ and $1/2^m \to 0$ as $m \to \infty$, both $(x_n - x_m)$ and $(y_n - y_m) \to 0$ as $m, n \to \infty$, and our assertion is proved. Now

$$|x_n - y_n| = \left|\left(2 - \frac{1}{2^n}\right) - \left(2 + \frac{1}{2^n}\right)\right|$$

$$= \left|-\frac{1}{2^n} - \frac{1}{2^n}\right| = \left|(-1)\left(\frac{1}{2^n} + \frac{1}{2^n}\right)\right|$$

$$= |-1|\left|\frac{2}{2^n}\right| = \frac{1}{2^{n-1}},$$

and since $1/2^{n-1} \to 0$ as $n \to \infty$, the sequence $x - y$ is a null sequence, that is, $x\mathcal{R}y$.

The relation $\mathcal{R}$ on C is an equivalence relation:

(i) $\mathcal{R}$ is *reflexive*, for if $x \in C$, then $x - x$ is the sequence

$$x_1 - x_1, x_2 - x_2, x_3 - x_3, \dots \quad \text{or} \quad 0, 0, 0, \dots;$$

hence $x\mathcal{R}x$.

(ii) $\mathcal{R}$ is *symmetric*, for if $x, y \in C$ and $x\mathcal{R}y$, then

$$(x_n - y_n) \to 0 \qquad \text{as} \qquad n \to \infty,$$

$$|y_n - x_n| = |-x_n + y_n| = |(-1)(x_n - y_n)|$$

$$= |-1| \, |x_n - y_n| = |x_n - y_n|.$$

Thus $(y_n - x_n) \to 0$ as $n \to \infty$ and $y\mathcal{R}x$.

(iii) $\mathcal{R}$ is *transitive*. To prove this, take any sequences $x, y, z \in C$ such that $x\mathcal{R}y$ and $y\mathcal{R}z$. Then

$$(x_n - y_n) \to 0 \quad \text{and} \quad (y_n - z_n) \to 0 \quad \text{as} \quad n \to \infty$$

But

$$0 \leq |x_n - z_n| = |(x_n - y_n) + (y_n - z_n)| \leq |x_n - y_n| + |y_n - z_n| \text{ (why?)},$$

so that $(x_n - z_n) \to 0$ as $n \to \infty$ and $x \mathfrak{R} z$.

Now that we have shown $\mathfrak{R}$ to be an equivalence relation on the set C of C-sequences, we can use $\mathfrak{R}$ to partition C as follows: For each element $t \in C$ let C_t be the subset of C containing those elements, and only those elements, $s \in C$ such that $s \mathfrak{R} t$. We thus have a family of subsets, a partition,

$$\mathfrak{C} = \{C_u, C_v, C_w, C_x, C_y, \ldots\},$$

where C_u is the set of all elements (C-sequences) $s \in C$ such that $s \mathfrak{R} u$, C_v is the set of all elements $s \in C$ such that $s \mathfrak{R} v$, and so on. We are aware that a subset belonging to the partition $\mathfrak{C}$ may appear in $\mathfrak{C}$'s membership list more than once, possibly infinitely many times.

If, for example, u and v are C-sequences and $u - v$ is a null sequence, so that $u \mathfrak{R} v$, then C_u and C_v will both appear in $\mathfrak{C}$'s membership list even though $C_u = C_v$. We will from now on assume that *any two* equivalence sets in the partition

$$\mathfrak{C} = \{C_u, C_v, C_w, C_x, C_y, \ldots\}$$

are *distinct* and therefore *disjoint*; in other words, we assume that all duplicates of $C_u, C_v, C_w, \ldots$ have been removed. No harm comes from doing this, for by eliminating duplications the membership of $\mathfrak{C}$ remains the same. The set $\mathfrak{C}$ that we now have will be called the *reals* and will eventually become our real number system. For convenience we will henceforth denote the reals by capital letters $U, V, W, X, \ldots$, remembering at all times that these capital letters represent *disjoint* subsets of the set C of all C-sequences of rational numbers and that they are the elements of $\mathfrak{C}$, the partition of C induced by the equivalence relation $\mathfrak{R}$.

We now define operations $+$ and $\cdot$ on $\mathfrak{C}$, to be called addition and multiplication.

DEFINITION. *Let U and V be any two elements of $\mathfrak{C}$. Select any element $u \in U$ and any element $v \in V$. Then we define $U + V = W$, where W is that element of $\mathfrak{C}$ such that $u + v \in W$, and we define $U \cdot V = X$, where X is that element of $\mathfrak{C}$ such that $u \cdot v \in X$.*

We have proved that the sum of any two C-sequences is always a C-sequence. Hence $u + v$ is a unique C-sequence and, for a particular u and v, a unique real W is determined, since only *one* member of the partition $\mathfrak{C}$ contains $u + v$. However, the definition for addition tells us to select *any* element $u \in U$ and *any* element $v \in V$. We now ask: If we select *some other* element $u' \in U$ and *some other* element $v' \in V$, will it still be true that $u' + v' \in W$? That is, is our addition operation well defined? To show that it is, we must show that

$$(u' + v')\Re(u + v)$$

or, what is the same thing, that

$$(u' + v') - (u + v)$$

is a null sequence. This is easily shown by considering

$$|(u'_n + v'_n) - (u_n + v_n)|$$
$$= |(u'_n - u_n) + (v'_n - v_n)| \le |u'_n - u_n| + |v'_n - v_n|.$$

Since u' and u both are elements of U, then $u'\Re u$, so that $u' - u$ is a null sequence and $(u'_n - u_n) \to 0$; by similar reasoning $(v'_n - v_n) \to 0$. Therefore

$$|(u'_n + v'_n) - (u_n + v_n)|$$

is squeezed toward 0 as $n \to \infty$ and

$$(u' + v') - (u + v)$$

is a null sequence and $u' + v' \in W$, as we wished to show. To show that multiplication is well defined we select *some other* element $u' \in U$ and *some other* element $v' \in V$ and show that $u' \cdot v'\Re u \cdot v$, or that $u' \cdot v' - u \cdot v$ is a null sequence. This is done by considering

$$|u'_n \cdot v'_n - u_n \cdot v_n| = |(u'_n \cdot v'_n - u'_n \cdot v_n) + (u'_n \cdot v_n - u_n \cdot v_n)|$$
$$= |u'_n(v'_n - v_n) + v_n(u'_n - u_n)| \le |u'_n(v'_n - v_n)| + |v_n(u'_n - u_n)|$$
$$= |u'_n| \cdot |v'_n - v_n| + |v_n| \cdot |u'_n - u_n| \le A|v'_n - v_n| + B|u'_n - u_n|,$$

where A and B are *bounds* for the sequences u' and v. Now since $u' \in U$, then $u'\Re u$, so that $u' - u$ is a null sequence and $(u'_n - u_n) \to 0$; by a similar argument $(v'_n - v_n) \to 0$. Thus, since

$$0 \le |u'_n \cdot v'_n - u_n \cdot v_n| \le A|v'_n - v_n| + B|u'_n - u_n|,$$

$(u'_n \cdot v'_n - u_n \cdot v_n) \to 0$ as $n \to \infty$ and $u' \cdot v' - u \cdot v$ is a null sequence, as we wished to show.

We now would like to know whether addition and multiplication of reals is commutative and associative and if multiplication is distributive with respect to addition. To answer our own question, we merely ask ourselves what we would do if we were given any reals U, V, W and were required to find $U + V$, $U \cdot V$, $(U + V) + W$, $(U \cdot V) \cdot W$, and

$U \cdot (V + W)$. We would first select any C-sequences u, v, w such that $u \in U$, $v \in V$, and $w \in W$. Then, using our definitions for addition and multiplication of reals, we would immediately see, since the commutative, associative, and distributive laws hold for rationals, that

$$u + v = v + u, \quad u \cdot v = v \cdot u, \quad (u + v) + w = u + (v + w),$$
$$(u \cdot v) \cdot w = u \cdot (v \cdot w), \quad u \cdot (v + w) = u \cdot v + u \cdot w,$$

and that, consequently, these same laws will hold for the reals. Our set $\mathcal{C}$ of reals together with the operations $+$ and $\cdot$ is a genuine number system and we may call the elements U, V, W, $X, \ldots$ belonging to $\mathcal{C}$ *real numbers* from now on.

Something that hasn't been specifically mentioned and which should be emphasized is the fact that *every null sequence is a C-sequence*. To show this, let x_1, x_2, x_3, $x_4, \ldots$ be *any* null sequence. Then, of course, we know that $x_n \to 0$ as $n \to \infty$. Thus,

$$0 \leq |x_n - x_m| \leq |x_n| + |x_m|,$$

and since
$$x_n \to 0 \quad \text{as} \quad n \to \infty \quad \text{and} \quad x_m \to 0 \quad \text{as} \quad m \to \infty,$$
we see that
$$(x_n - x_m) \to 0 \quad \text{as} \quad m, n \to \infty,$$

proving that x_1, x_2, x_3, $x_4, \ldots$ is a C-sequence. As a result, we conclude that one of our equivalence sets (real numbers) belonging to $\mathcal{C}$ has all the null sequences as members. We shall designate this real number by Z and show that it is the unique additive identity (the zero element) in our real number system; that is, $U + Z = U$ for any $U \in \mathcal{C}$. This is certainly true, for since addition is well defined, we may select *any* element $u \in U$ and *the particular* element $0, 0, 0, 0, \ldots$ (the trivial null sequence) from Z, and since $u_1 + 0$, $u_2 + 0$, $u_3 + 0, \ldots$ is merely the sequence $u \in U$, then $U + Z = U$, as we wished to show.

Now that we know that the real number system has a unique additive identity, we may ask if every real number has an additive inverse. We may answer this by noting that if $u : u_1, u_2, u_3, \ldots$ is any C-sequence of rational numbers, then $-u : -u_1, -u_2, -u_3, \ldots$ is also a C-sequence (prove it!). Thus, if $u \in U$, we will denote the real number (set) containing $-u$ by $-U$ and we will have $U + (-U) = Z$, by definition of addition. Now observe that we have a set $\mathcal{C}$ (the reals) with a well-defined operation $+$ defined on $\mathcal{C}$, that this operation is associative, that there is a unique additive identity element $Z \in \mathcal{C}$, and that for each element $U \in \mathcal{C}$ there is an element $-U \in \mathcal{C}$ such that $U + (-U) = Z$. In short, the set $\mathcal{C}$ is a group with respect to addition and we immediately know that each real number has a *unique* additive inverse.

We can now define the operation subtraction on $\mathbb{C}$ (the reals).

DEFINITION. If U and V are any real numbers, we define

$$U - V = U + (-V).$$

Note that $U + (-V)$ is that real number which added to V will yield U.

$$[U + (-V)] + V = U + [(-V) + V] = U + Z = U.$$

The real number system also has a unique multiplicative identity which we designate by I. You have probably guessed by now that I is the real number (set) containing the trivial C-sequence 1, 1, 1, 1, ... If we wish to find $U \cdot I$ we select *any* element $u \in U$ and the *particular* element 1, 1, 1, 1, ... from I. Thus, since $u_1 \cdot 1, u_2 \cdot 1, u_3 \cdot 1, \ldots$ is merely the C-sequence $u \in U$, then $U \cdot I = U$, as we wished to prove.

Now that the reals have been shown to have a multiplicative identity, we may fairly ask whether each real (other than Z) has a multiplicative inverse. We *suspect* that for each real number $U \neq Z$, there exists a real number U^{-1} such that $U \cdot U^{-1} = I$, but in order to confirm our suspicion, we will first need to discuss *subsequences* of rational numbers. We begin by saying that a subsequence of a given sequence is a sequence that may be constructed by selecting a first term, a second term, a third term, ... from the terms of the given sequence and preserving their relative order. Consider, for example, the sequence of positive integers

$$P: 1, 2, 3, 4, 5, \ldots$$

The sequence

$$Q: 1, 3, 5, 7, 9, \ldots$$

of odd positive integers is clearly a subsequence of P, for each of its terms is a term of P and, furthermore, any two terms of Q have the same relative order that they have in P; that is, if h and k represent any terms of Q, then h is to the left of k in Q only if h is to the left of k in P. The sequence

$$R: 2, 4, 6, 8, 10, \ldots$$

of even positive integers and the sequence

$$S: 1, 2, 3, 5, 7, 11, 13, \ldots$$

of prime positive integers are further examples of subsequences of P. If we have a given sequence x: $x_1, x_2, x_3, \ldots$ and we wish to construct a subsequence of x, it is not necessary to have a rule for selecting the terms of the subsequence; we can select the terms at random just as long as we preserve their relative order.

Let us imagine that we have a wheel of chance with the numbers 1, 2, 3, 4, . . . , r marked evenly around its circumference (r is a fixed but arbitrary positive integer) and let us construct a subsequence of x by spinning our imaginary wheel. Suppose that the wheel stops at 4 on the *first* spin and we select x_4 as the *first* term of the subsequence, the wheel stops at 7 on the *second* spin and we select x_{11} as the *second* term of our subsequence, the wheel stops at 21 on the *third* spin and we select x_{32} as the *third* term of our subsequence, and so on. We see that this process, if continued indefinitely, will actually yield a subsequence of x_1, x_2, x_3, x_4, . . . , in our case the sequence

$$x_4, x_{11}, x_{32}, \ldots$$

If x_1, x_2, x_3, x_4, . . . is a given sequence, it is common practice in mathematical writing to let

$$x_{n_1}, x_{n_2}, x_{n_3}, \ldots$$

denote a general subsequence, where n_1 represents the *number* of the *first* term selected, n_2 represents the *number* of the *second* term selected, and so on. In our particular imaginary subsequence we had

$$n_1 = 4, n_2 = n_1 + 7 = 11, n_3 = n_2 + 21 = 32, \ldots$$

In general, n_1, n_2, n_3, n_4, . . . are positive integers such that

$$n_1 < n_2 < n_3 < n_4 < \ldots \qquad \text{and} \qquad n_1 \geq 1, n_2 \geq 2, n_3 \geq 3, \ldots$$

Below we display a sequence alongside one of its subsequences.

$$x_1, x_2, \ldots, x_{n_1-1}, x_{n_1}, x_{n_1+1}, \ldots, x_{n_2-1}, x_{n_2}, x_{n_2+1}, \ldots, x_{n_3-1}, x_{n_3}, \ldots$$
$$x_{n_1}, \qquad\qquad x_{n_2}, \qquad\qquad x_{n_3}, \ldots$$

We now prove a theorem about subsequences.

THEOREM I. *Let x*: x_1, x_2, x_3, . . . *be any C-sequence of rational numbers and let x'*: x_{n_1}, x_{n_2}, x_{n_3}, . . . *be any subsequence of x. Then x' is a C-sequence and $x - x'$ is a null sequence.*

Proof. Since x is a C-sequence,

$$(x_p - x_q) \to 0 \quad \text{as} \quad p, q \to \infty.$$

This, of course, is the same as saying that corresponding to any positive integer k a positive integer T can be found such that

$$|x_p - x_q| < 10^{-k}$$

whenever $p, q > T$. Since $n_p \geq p$ and $n_q \geq q$, it follows that $n_p, n_q > T$

whenever p, $q > T$ and that

$$|x_{n_p} - x_{n_q}| < 10^{-k},$$

thus proving that x' is a C-sequence.

To prove that $x - x'$ is a null sequence we merely observe once more that

$$|x_p - x_q| < 10^{-k}$$

for any integers p, $q > T$. In particular, since $n_p \geq p$ and $n_p > T$ whenever $p > T$, we can take $q = n_p$. We therefore will have

$$|x_p - x_{n_p}| < 10^{-k}$$

whenever $p > T$, proving that

$$(x_p - x_{n_p}) \to 0 \quad \text{as} \quad p \to \infty,$$

and $x - x'$ is a null sequence, as we wished to show.

Now that Theorem I is proved we can rephrase it: If U is any real number and $x \in U$, then any subsequence of x also is an element of U. An immediate *corollary* of this theorem is

COROLLARY. If $x : x_1, x_2, x_3, \ldots$ is any non-null C-sequence of rational numbers, then x has at most a finite number of zero terms.

Proof. Assume for the sake of argument that x has infinitely many zero terms. Then the trivial null sequence $0, 0, 0, 0, \ldots$ is a subsequence of x. But, by Theorem I, $x_1 - 0$, $x_2 - 0$, $x_3 - 0$, $\ldots$ (that is, x itself) is a null sequence. This contradicts our hypothesis that x is a non-null sequence and we conclude that our assumption is false; x has at most a finite number of zero terms.

As a result of this corollary we are assured that if a non-null sequence $x_1, x_2, x_3, \ldots$ has any zero terms, we can start with x_1 and, by moving to the right, actually *arrive at* and *go beyond* the *last* zero term. Moreover, we can always construct a subsequence of $x_1, x_2, x_3, \ldots$ that has no zero terms, for example by selecting only terms beyond the last zero term or by merely selecting all the nonzero terms. For the moment we are interested in non-null C-sequences that have no zero terms and we prove the following theorem:

THEOREM II. Let $x : x_1, x_2, x_3, \ldots$ be a non-null C-sequence having no zero terms. Then there exists a positive rational A such that $A < |x_n|$ for all values of the positive integer n.

Proof. Let us assume that the theorem is false; that is, if we take any positive rational, as small as we please, we can always find a term of the sequence whose absolute value is smaller. If our assumption is correct, we can start with x_1 and, moving out in the sequence $x : x_1, x_2, x_3, \ldots$; find a *first* term x_{n_1} such that $|x_{n_1}| < 10^{-1}$; we can move beyond x_{n_1} and find a *second* term x_{n_2} such that $|x_{n_2}| < |x_{n_1}|$ and $|x_{n_2}| < 10^{-2}$; we can move beyond x_{n_2} and find a *third* term x_{n_3} such that $|x_{n_3}| < |x_{n_2}|$ and $|x_{n_3}| < 10^{-3}$, and so on. By continuing this selection process indefinitely a subsequence $x' : x_{n_1}, x_{n_2}, x_{n_3}, \ldots$ may be obtained which is obviously a null sequence and is an element of the real number Z. But, by the previous theorem, $x - x'$ is a null sequence, so that $x \Re x'$ and $x \in Z$; that is, x is a null sequence also. This conclusion contradicts our hypothesis and our assumption is necessarily false; the theorem is therefore true.

Another theorem we can now prove is:

THEOREM III. If $x : x_1, x_2, x_3, \ldots$ is a C-sequence with no zero terms, then the sequence $y : y_1, y_2, y_3, \ldots$, where $y_n = 1/x_n$, is also a C-sequence.

Proof. To prove that $y_1, y_2, y_3, \ldots$ is a C-sequence we must show that

$$(y_n - y_m) \to 0 \quad \text{as} \quad m, n \to \infty.$$

Thus we shall consider $|y_n - y_m|$ and see what happens to it as $m, n \to \infty$. We first observe that

$$|y_n - y_m| = \left| \frac{1}{x_n} - \frac{1}{x_m} \right| = \left| \frac{x_m - x_n}{x_n \cdot x_m} \right| = \frac{|x_n - x_m|}{|x_n| \cdot |x_m|}$$
$$= \frac{1}{|x_n| \cdot |x_m|} \cdot |x_n - x_m|.$$

Since x is a C-sequence with no zero terms, there is a positive rational A such that $A < |x_n|$ for every positive integer n (Theorem II). Thus

$$\frac{1}{|x_n|} < \frac{1}{A}, \quad \frac{1}{|x_m|} < \frac{1}{A} \quad \text{(Why?)},$$

and

$$\frac{1}{|x_n| \cdot |x_m|} < \frac{1}{A^2}. \quad \text{(Why?)},$$

Therefore we have

$$0 \le |y_n - y_m| < \frac{1}{A^2} |x_n - x_m|,$$

and, since $(x_n - x_m) \to 0$ as $m, n \to \infty$, $|y_n - y_m|$ is squeezed toward 0 as $m, n \to \infty$, and $y : y_1, y_2, y_3, \ldots$ is a C-sequence, as we wished to show.

The theorems we have just finished proving make it possible for us to show now that any nonzero real number has a multiplicative inverse. To this end, let U be a nonzero real number ($U \neq Z$) and select a C-sequence $x : x_1, x_2, x_3, \ldots$ having no zero terms from U; this is always possible, as we have already proved. We assert that the real number V which contains the C-sequence $1/x_1, 1/x_2, 1/x_3, \ldots$ is the multiplicative inverse of U; that is, $U \cdot V = I$. To prove this assertion we shall need to actually multiply U and V. To help us, we recall that multiplication of reals is *well-defined;* that is, our product $U \cdot V$ does not depend on what elements are selected from the memberships of U and V. Therefore we select the *particular* element $x : x_1, x_2, x_3, \ldots$ from U and the *particular* element $y : y_1, y_2, y_3, \ldots$ ($y_n = 1/x_n$) from V. Since $x \cdot y$ is the trivial C-sequence $1, 1, 1, 1, \ldots$ belonging to the real number I, we have, by definition of multiplication, $U \cdot V = I$. Using the notation U^{-1} (U inverse) to denote the multiplicative inverse of any nonzero real number U, we define division as follows.

DEFINITION. Let X and Y be any real numbers such that $Y \neq Z$. Then, we define $X \div Y = X \cdot Y^{-1}$.

Note that the quotient $X \cdot Y^{-1}$ is that real number which multiplied by Y (the divisor) will yield X (the dividend).

We should observe that the set of nonzero reals is a group with respect to multiplication, and consequently the multiplicative cancellation property holds in the reals: if $U \cdot V = U \cdot W$ and $U \neq Z$, then $V = W$. Furthermore, the additive and multiplicative identity elements are distinct ($Z \neq I$) so that $\{\mathcal{C}; +, \cdot\}$ is an integral domain. Therefore it is clear (Section 8–8) that the number system $\{\mathcal{C}; +, \cdot\}$ is a field. We must next turn our attention to showing in what sense the rationals are a subfield of the reals.

9–6 The infinite decimals. In Section 9–3 we discussed the possibility of having a number system whose elements are the *infinite decimals*, both repeating and nonrepeating. We are now in position to show that there is a 1 : 1 correspondence between the set $\mathcal{C}$ of real numbers and the set of all infinite decimals. It will take a little work to do this, and we start out by showing that *each infinite decimal corresponds to a unique real number.*

Proof. Let
$$10^r \times a_0.a_1a_2a_3a_4 \ldots$$

be any positive infinite decimal and note that this infinite decimal corresponds to a unique sequence $x : x_1, x_2, x_3, x_4, \ldots$ of rationals, where

$$x_1 = 10^r \times a_0.a_1, \quad x_2 = 10^r \times a_0.a_1a_2, \quad x_3 = 10^r \times a_0.a_1a_2a_3, \ldots$$

and, in general, $x_n = 10^r \times a_0.a_1a_2 \ldots a_n$. We assert that x is a C-sequence. To prove this we consider $|x_n - x_m|$ and show that for any positive integer k there exists a corresponding positive integer T such that

$$|x_n - x_m| < 10^{-k}$$

whenever $m, n > T$. Since $|x_n - x_m| = 0$ if $m = n$, we may assume that $m \neq n$ and, for definiteness, we shall suppose that $m > n$. In this case, then, we may write

$$|x_n - x_m| = |10^r \times a_0.a_1a_2 \ldots a_n - 10^r \times a_0.a_1a_2 \ldots a_n \ldots a_m|$$

$$= 10^r \times |a_0.a_1a_2 \ldots a_n - a_0.a_1a_2 \ldots a_n \ldots a_m|$$

$$\overbrace{}^{n \ 0\text{'s}}$$

$$= 10^r \times 0.00 \ldots 0a_{n+1} \ldots a_m$$

$$= 10^r \times 10^{-(n+1)} \times a_{n+1}.a_{n+2} \ldots a_m.$$

Now, each of the digits $a_{n+1}, a_{n+2}, \ldots, a_m$ is an integer of the set $\{0, 1, 2, \ldots, 9\}$ and $a_{n+1}.a_{n+2} \ldots a_m < 10$ so that

$$|x_n - x_m| < 10^r \times 10^{-(n+1)} \times 10 = 10^{r-n}.$$

Thus, for a given positive integer k and corresponding rational 10^{-k}, we can make $10^{r-n} < 10^{-k}$, provided that we take $n > k + r$ (why?). It therefore follows that we may take $T = k + r$ and say that $|x_n - x_m| < 10^{-k}$ whenever $m > n > T$. Since a similar argument could have been made had we supposed $n > m$, it is clear that

$$(x_n - x_m) \to 0 \quad \text{as} \quad m, n \to \infty,$$

and our assertion is proved. Had we taken a negative infinite decimal $-10^r \times a_0.a_1a_2a_3 \ldots$, we would have considered the sequence $-x$: $-x_1, -x_2, -x_3, \ldots$ and, since

$$|(-x_n) - (-x_m)| = |-x_n + x_m|$$

$$= |(-1)(x_n - x_m)|$$

$$= |-1| \, |x_n - x_m|$$

$$= |x_n - x_m|,$$

we would have reached the same conclusion. Moreover, the infinite decimal $0.00000\overline{0}$ corresponds to the trivial null sequence 0.0, 0.00, 0.000,

... or, simply, 0, 0, 0, 0, ... Thus any infinite decimal

$$\pm 10^r \times a_0.a_1 a_2 a_3 \ldots$$

corresponds to a C-sequence

$$\pm 10^r \times a_0.a_1, \quad \pm 10^r \times a_0.a_1 a_2, \quad \pm 10^r \times a_0.a_1 a_2 a_3, \ldots,$$

and we shall call this sequence the sequence *generated* by the infinite decimal. Thus we *define* a mapping, β, which associates each infinite decimal with the real number containing its generated sequence. In short, if D is the set of all infinite decimals, we have defined a mapping β of D *into* $\mathcal{C}$. (Remember that there is no infinite decimal in D each of whose digits from some point on is a 9.)

To show that β is a 1 : 1 mapping of D onto $\mathcal{C}$ we must prove (i) that β maps distinct infinite decimals into distinct real numbers, and (ii) that β is a mapping of D *onto* $\mathcal{C}$. We now state (i) as a *theorem*; Let

$$x : x_1, x_2, x_3, \ldots \quad \text{and} \quad y : y_1, y_2, y_3, \ldots$$

be generated sequences such that $x - y \in Z$. Then $x_n = y_n$ for all values of the positive integer n. This theorem may be paraphrased: "If P and Q are infinite decimals corresponding to the same real number, then their generated sequences are term-by-term identical and $P = Q$."

Proof. It will be sufficient to prove this theorem for sequences generated by *positive* infinite decimals (why?). Let us assume that the sequences x and y are *not* term-by-term identical. Then, if this assumption is correct, there is a *first* pair of corresponding terms, say x_p and y_p, which are not equal; in other words,

$$x_p = 10^r \times a_0.a_1 a_2 \ldots a_{p-1} a_p \quad \text{and} \quad y_p = 10^r \times a_0.a_1 a_2 \ldots a_{p-1} b_p.$$

For definiteness, let us suppose that $a_p > b_p$, so that $a_p = b_p + m$, where m is one of the integers $\{1, 2, 3, \ldots, 8, 9\}$.

Case 1 ($m \geq 2$). Consider the difference between any two corresponding terms which lie beyond x_p and y_p, say $x_{p+q} - y_{p+q}$. Examining this difference, we find

$$x_{p+q} - y_{p+q}$$

$$= 10^r \times a_0.a_1 a_2 \ldots a_p \ldots a_{p+q} - 10^r \times a_0.a_1 a_2 \ldots b_p \ldots b_{p+q}$$

$$= 10^r \times (0.00 \ldots a_p \ldots a_{p+q} - 0.00 \ldots b_p \ldots b_{p+q}).$$

We do not know the digits beyond a_p and b_p, but we do know that if the difference $x_{p+q} - y_{p+q}$ is to be as small as possible, we must have

$$a_{p+1} = a_{p+2} = \ldots = a_{p+q} = 0 \quad \text{and} \quad b_{p+1} = b_{p+2} = \ldots = b_{p+q} = 9.$$

Whether this is the case or not, we will have

$$x_{p+q} - y_{p+q} \geq 10^r \times (0.00 \ldots \overline{m - 1}00 \ldots 01)$$

$$= 10^r \times (10^{-p} \times \overline{m - 1}.00 \ldots 01)$$

$$= 10^{r-p} \times \overline{m - 1}.00 \ldots 01,$$

and since $m - 1 \geq 1$,

$$x_{p+q} - y_{p+q} \geq 10^{r-p} \times 1.\overbrace{00 \ldots 01}^{(q-1) \text{ 0's}} > 10^{r-p}.$$

In this case $(m \geq 2)$, then, we see that all the terms in the sequence $x - y$ beyond the pth term are *greater* than the positive rational 10^{r-p} and $x - y$ is *not* a null sequence.

Case 2 ($m = 1$). We again consider the difference $x_{p+q} - y_{p+q}$, where q is a *certain* positive integer to be chosen in a special way. We examine the infinite decimal Q which generates $y : y_1, y_2, y_3, \ldots$:

$$Q = 10^r \times a_0.a_1a_2 \ldots a_{p-1}b_p b_{p+1} b_{p+2} \ldots .$$

We know that to the right of the digit b_p there are *at most a finite number* of consecutive 9's; that is, if we go beyond b_p we will encounter a *first* digit which is *not* a 9. We denote this digit by b_{p+q}. Thus q is the number of decimal places to the right of b_p that the first digit not equal to 9 occurs. Hence,

$$x_{p+q} - y_{p+q} = 10^r \times (a_0.a_1a_2 \ldots a_p \ldots a_{p+q} - a_0.a_1a_2 \ldots b_p \ldots b_{p+q})$$

$$= 10^r \times (0.00 \ldots a_p \ldots a_{p+q} - 0.00 \ldots b_p \ldots b_{p+q})$$

$$= 10^{r-p} \times (a_p.a_{p+1} \ldots a_{p+q} - b_p.b_{p+1} \ldots b_{p+q}).$$

Again, we do not know the digits beyond a_p and b_p, but we do know that if the difference $x_{p+q} - y_{p+q}$ is to be as small as possible, we must have

$$a_{p+1} = a_{p+2} = \ldots = a_{p+q} = 0,$$

$$b_{p+1} = b_{p+2} = \ldots = b_{p+q-1} = 9, \quad b_{p+q} = 8,$$

so that

$$x_{p+q} - y_{p+q} \geq 10^{r-p} \times (a_p.00\ldots0 - b_p.99\ldots8)$$

$$= 10^{r-p} \times (0.\overbrace{00\ldots0}^{(q-1)\ 0\text{'s}}2)$$

$$= 10^{r-p} \times (10^{-q} \times 2) = 10^{r-p-q} \times 2.$$

Now consider the difference $x_{p+q+s} - y_{p+q+s}$ for some positive integer s. This difference is as small as possible if

$$a_{p+1} = a_{p+2} = \ldots = a_{p+q} = a_{p+q+1} = \ldots = a_{p+q+s} = 0,$$

$$b_{p+1} = b_{p+2} = \ldots = b_{p+q-1} = b_{p+q+1} = \ldots = b_{p+q+s} = 9,$$

$$b_{p+q} = 8.$$

Therefore

$$x_{p+q+s} - y_{p+q+s}$$

$$\geq 10^r \times (a_0.a_1a_2\ldots a_p\overbrace{00\ldots0\ldots0}^{(q+s)\ 0\text{'s}} - a_0.a_1a_2\ldots b_p99\ldots89\ldots.$$

$$= 10^r \times (0.\overbrace{00\ldots01}^{(p+q-1)\ 0\text{'s}}00\ldots01)$$

$$= 10^r \times 10^{-(p+q)} \times 1.\overbrace{00\ldots01}^{(s-1)\ 0\text{'s}}$$

$$= 10^{r-p-q} \times 1.00\ldots01 > 10^{r-p-q}.$$

In this case ($m = 1$), then, all the terms in the sequence $x - y$ beyond the $(p+q)$th term are *greater* than the positive rational 10^{r-p-q} and $x - y$ is obviously *not* a null sequence.

In both Case 1 and Case 2 our assumption that x and y are not term-by-term identical has led us to the conclusion that $x - y$ is *not* a null sequence. Since our hypothesis is contradicted we know that our theorem is true.

We have just proved that whenever two infinite decimals correspond to the same real number, they must necessarily be equal; this, of course, is the same as proving that *distinct* decimals correspond to *distinct* reals. It now remains to prove (ii): if U is any real number, there exists an infinite decimal P such that $\beta P = U$.

Proof. We have asserted that an infinite decimal P *exists* such that $\beta P = U$. To prove this, we must show that a generated sequence $y : y_1,$ $y_2, y_3, \ldots$ *exists* such that $y \in U$. It is important to understand that we are not obliged to *exhibit* a generated sequence y but, rather, to merely prove the *existence* of one. We start by selecting a sequence $x : x_1, x_2,$ $x_3, \ldots$ from U and, since x is a C-sequence of *rational numbers*, there is a term x_{n_1} such that

$$|x_{n_1} - x_{m+n_1}| < 10^{-1},$$

there is a term x_{n_2} such that

$$|x_{n_2} - x_{m+n_2}| < 10^{-2},$$

there is a term x_{n_3} such that

$$|x_{n_3} - x_{m+n_3}| < 10^{-3},$$

and so on, for any value of the positive integer m. Therefore, the subsequence

$$x' : x_{n_1}, x_{n_2}, x_{n_3}, x_{n_4}, \ldots$$

has the property that $|x_{n_k} - x_{n_{m+k}}| < 10^{-k}$ for all values of the positive integer m. Now let each term $x_{n_1}, x_{n_2}, x_{n_3}, \ldots$ be expressed as an infinite decimal. Then, since

$$|x_{n_1} - x_{n_{m+1}}| < 10^{-1}$$

for *all* values of the positive integer m, there are *infinitely many* terms of x' having a common part $a_p \ldots a_1 a_0 . b_1$. Although any two terms of x' differ by less than 10^{-1}, we cannot say that *all* terms of x' have the same digits to the left of the 100ths place. For example, one may have a finite sequence of consecutive 0's to the left of the 100ths place while another has a finite sequence of corresponding 9's. In any case, we can say that *infinitely many* terms of x' have a common part $y_1 = a_p \ldots a_1 a_0 . b_1$ to the left of the 100ths place. Then, since $|x_{n_2} - x_{n_{m+2}}| < 10^{-2}$ for *all* values of the positive integer m, of the infinitely many terms of x' having a common part y_1, there are infinitely many having the same 100ths digit b_2. Thus, letting

$$y_2 = y_1 + 0.0b_2 = a_p \ldots a_1 a_0 . b_1 b_2,$$

we can say that there are infinitely many terms of x' having common part y_2 to the left of the 1000ths place. Continuing, since $|x_{n_3} - x_{n_{m+3}}| < 10^{-3}$ for all values of the positive integer m, of the infinitely many terms of x' having common part y_2, there are infinitely many having the same 1000ths digit b_3. Thus, letting

$$y_3 = y_2 + 0.00b_3 = a_p \ldots a_1 a_0 . b_1 b_2 b_3,$$

we can say that there are infinitely many terms of x' having common part y_3 to the left of the 10,000ths place. This argument, of course, may be continued indefinitely to establish the *existence* of a generated sequence $y : y_1, y_2, y_3, \ldots$ and the infinite decimal $P = a_p \ldots a_1 a_0 . b_1 b_2 b_3 b_4 \ldots$ which generates it. We assert that $\beta P = U$, and to prove this we must show that $y \in U$. We know that there is a *first* term of $x' : x_{n_1}, x_{n_2}, x_{n_3}, \ldots$, say y_1', such that $|y_1' - y_1| < 10^{-1}$, there is a *second* term of x', say y_2', such that $|y_2' - y_2| < 10^{-2}$, there is a *third* term of x', say y_3', such that $|y_3' - y_3| < 10^{-3}$, and so on. That is, the subsequence x' has a subsequence $y_1', y_2', y_3', \ldots$ such that $(y_h' - y_h) \to 0$ as $h \to \infty$. We now recall our equivalence relation on the set C of C-sequences: If $u : u_1, u_2, u_3, \ldots$ and $v : v_1, v_2, v_3, \ldots$ are C-sequences, then $u \mathcal{R} v$ if, and only if, $(u_n - v_n) \to 0$ as $n \to \infty$. In particular, $\mathcal{R}$ is *transitive*. Therefore, since $(y_h' - y_h) \to 0$, $(x_{n_h} - y_h') \to 0$, and $(x_h - x_{n_h}) \to 0$, it follows that $(x_h - y_h) \to 0$ as $h \to \infty$ and the generated sequence $y : y_1, y_2, y_3, \ldots$ belongs to U, as we wished to show.

We have now shown what we set out to show: there is a $1 : 1$ correspondence between the set of all real numbers and the set of all infinite decimals. In particular, the real number Z corresponds to the infinite decimal $0.0000 \ldots$ and the real number I corresponds to the infinite decimal $1.0000 \ldots$ We now can define operations on the set of infinite decimals. We define addition and multiplication so that our $1 : 1$ mapping β is an *isomorphism* with respect to each operation and, of course, the commutative, associative, and distributive laws will automatically hold (Problem 4, Exercise Group 5–3). In fact, the system $\{D; +, \cdot\}$ is the *real number field* for we use the infinite decimals *as symbols for the real numbers* just as we earlier used the fractions as symbols for the rational numbers.

To actually add two infinite decimals d_1 and d_2, we add their corresponding real numbers, say U and V. Then, if $U + V = W$, we choose the generated sequence belonging to W and take the corresponding infinite decimal as the sum of d_1 and d_2. To multiply d_1 and d_2 we multiply the corresponding real numbers U and V. Then, if $U \cdot V = X$, we choose the generated sequence belonging to X and take the corresponding infinite decimal as the product of d_1 and d_2. Adding and multiplying infinite decimals are infinite processes and, of course, in actual practice these processes can be started but never completed unless both infinite decimals are rationals. One may ask, then, what practical use such operations have. To answer this question we shall need first to define an order relation on the infinite decimals.

An infinite decimal is said to be *positive* if, and only if, the *first* term of its generated sequence is positive. For example, the infinite decimals $2709.3275449715 \ldots$ and $0.000417535 \ldots$ are positive infinite decimals. The generated sequence of the first is

$$10^3 \times 2.7, \ 10^3 \times 2.70, \ 10^3 \times 2.709, \ 10^3 \times 2.7093, \ldots,$$

and the generated sequence of the second is

$$10^{-4} \times 4.1, \ 10^{-4} \times 4.17, \ 10^{-4} \times 4.175, \ 10^{-4} \times 4.1753, \ldots$$

Of course, if the first term of a generated sequence is positive, *all* terms are positive. It is now natural to define a negative infinite decimal as one whose generated sequence has a negative rational as its first term. Thus $-73.28115483\ldots$ and $-0.000956127\ldots$ are negative infinite decimals whose generated sequences are

$$-10 \times 7.3, \ -10 \times 7.32, \ -10 \times 7.328, \ -10 \times 7.3281, \ldots$$

and

$$-10^{-4} \times 9.5, \ -10^{-4} \times 9.56, \ -10^{-4} \times 9.561, \ -10^{-4} \times 9.5612, \ldots$$

There is one and only one infinite decimal that is neither positive nor negative, namely $0.0000\ldots$, whose generated sequence is

$$0.0, \ 0.00, \ 0.000, \ 0.0000, \ldots$$

Naturally, we will designate this infinite decimal (real) simply by 0.

To define an order relation on the real numbers that is consistent with the order relation previously defined on the rationals we say simply this: A real number x is less than a real number $y(x < y)$ if, and only if, $y - x$ is positive ($y - x > 0$). It is often difficult to determine the generated sequence of $y - x$, but a practical way to tell whether $y - x$ is positive is to examine the corresponding terms of the generated sequences of x and y: find the *first* pair of corresponding terms that are unequal and if the term in x's sequence is less then the term in y's sequence, then $x < y$. To see how this works, examine $x = 29.3751\ldots$ and $y = 207.3852\ldots$, whose generated sequences are

$$10 \times 2.9, \ 10 \times 2.93, \ 10 \times 2.937, \ldots$$

and

$$10^2 \times 2.0, \ 10^2 \times 2.07, \ 10^2 \times 2.073, \ldots$$

It is clear that $10 \times 2.9 < 10^2 \times 2.0$; therefore $x < y$.

We return now to the question of how infinite decimals are actually added and multiplied in everyday arithmetic. The answer is *they are not*, except in the case where both are rational (repeating). According to the definition, to add $37.998352\ldots$ and $248.702094\ldots$ we would add the corresponding terms of their generated sequences, obtaining

$$10 \times 3.7 + 10^2 \times 2.4,$$
$$10 \times 3.79 + 10^2 \times 2.48,$$
$$10 \times 3.799 + 10^2 \times 2.487, \ldots$$

or

$$37 + 240, \; 37.9 + 248, \; 37.99 + 248.7, \ldots$$

The resulting sequence gives rise to a generated sequence which determines our sum. How do we *find* this sum? We actually don't *find* it; we simply write one decimal beneath the other, with the decimal points aligned, after first deleting the digits to the right of the 10ths place, or the 100ths place, or the 1000ths place, ..., depending on how precise we want our sum to be. In other words, we obtain *a rational sum* which differs from the *true* sum by as little as we please. If, in the above example, we desired a sum which differed from the true sum by not more than 10^{-2}, we would write

$$248.702094 \ldots$$
$$\underline{+37.998352} \ldots$$
$$286.700$$

and, rounding off to the nearest 100th, we would obtain 286.70. Here we have kept digits in the 1000ths place for safety. We observe that 286.70 is less than the true sum, while 286.71 would be greater than the true sum. At this point we merely remark that multiplication, subtraction, and division is done in a similar way and, of course, if we are dealing with real numbers which are rational, we may obtain *exact* answers by working with fractions instead of with infinite repeating decimals.

Much more could be said about performing operations with the reals. However, it has been assumed that the reader already knows "how"; our concern has been with "what" and "why."

EXERCISE GROUP 9–3

Note: $u, v, w, x, \ldots$ represent real numbers.

1. If u and v are positive, show that $u + v$ is positive.

2. Show that the relation $<$ on the real numbers is an order relation.

3. Let $<_1$ denote the order relation defined on the rationals (page 137) and let $<_2$ denote the order relation defined on the reals (page 182). If p/q and r/s are rationals, prove that $p/q <_1 r/s$ if, and only if, $p/q <_2 r/s$.

4. Show that if $u = v$, then $u + w = v + w$.

5. Show that if $u = v$, then $u \cdot w = v \cdot w$.

6. Show that if $u < v$, then $u + w < v + w$.

7. Show that if $0 < u$ and $0 < v$, then $0 < u \cdot v$.

8. Show that if $u < 0$ and $0 < v$, then $u \cdot v < 0$.

9. Show that if $u < v$ and $0 < w$, then $u \cdot w < v \cdot w$.

10. Show that if $u < v$ and $w < 0$, then $v \cdot w < u \cdot w$.

9–7 Countability. From our brief discussion of the infinite decimals we have learned that the repeating decimals (rationals) and the non-repeating decimals (irrationals) are intermingled; we would now like to know how they are intermingled and whether or not there are the "same number" of irrationals as rationals. We know that the rationals are dense (page 139); that is, between any two given rationals there is another rational. We ask if the irrationals are also dense; is there an irrational between any two given irrationals? Before trying to answer this question, let us observe that if an infinite decimal has no digits to the left of the decimal point we may attach any number of zeros; also, if an infinite decimal has one or more nonzero digits to the left of the decimal point, we may attach any number of zeros to the left of the first nonzero digit. For example, we may write .02750418 ... as 00.02750418 ... and we may write 507.40128 ... as 000507.40128 ... The point of this observation is that if two given infinite decimals have an unequal number of digits to the left of their respective decimal points, zeros may be attached so that the two decimals will *appear* to have the same number of digits to the left of their decimal points. The infinite decimals above will appear to have the same number of digits to the left of their decimal points if we write the first decimal as 000.02750418 ...

Now, to show that the irrationals are dense, we first take two positive irrationals d_1 and d_2 such that $d_1 < d_2$ and suppose that zeros are attached (if necessary) so that each decimal has the same number of digits to the left of its decimal point. If we now shift each decimal point to between the two leftmost digits and multiply each decimal by the same proper power of ten to compensate for these shifts, we will have our infinite nonrepeating decimals in the form

$$d_1 = 10^r \times a_0.a_1a_2a_3a_4 \ldots ,$$
$$d_2 = 10^r \times b_0.b_1b_2b_3b_4 \ldots$$

Since $d_1 < d_2$, there is a *first* pair of corresponding digits, say a_h and b_h, such that $a_h < b_h$. If $a_h + 2 \leq b_h$, the infinite decimal d_1 with the digit a_h replaced by $a_h + 1$ will be an infinite nonrepeating decimal (irrational) lying between d_1 and d_2 (why?). If, on the other hand, $a_h + 1 = b_h$, there is a *first* digit, say a_k, to the right of a_h such that $a_k \neq 9$ (why?) and the infinite decimal d_1 with the digit a_k replaced by $a_k + 1$ will be an infinite nonrepeating decimal (irrational) lying between d_1 and d_2. It is left as an exercise to prove by a similar argument that there is an irrational number lying between irrationals d_1 and d_2 if $d_1 < d_2$ and both are negative.

Finally, let us suppose that d_1 is negative and d_2 is positive, so that $d_1 < 0 < d_2$, and also suppose that $d_2 = 10^r \times b_0.b_1b_2b_3b_4 \ldots$ Then, since d_2 is positive, there is a *first* digit, say b_h, such that $b_h \neq 0$ and the decimal d_2 with the digit b_h replaced by $b_h - 1$ is a positive (why?) infinite nonrepeating decimal (irrational) less than d_2. By the transitivity of the $<$ relation, this irrational will lie between d_1 and d_2. We have proved that there is one irrational (hence infinitely many) between *any* two given irrationals; the irrationals are "everywhere dense."

Now that we know that the set of real numbers contains infinitely many rationals and also infinitely many irrationals, it might seem reasonable to guess that there are "as many" of one kind as the other contained in the set of reals. This is not the case, however; there are actually "many more" irrational numbers than rationals. To explain the meaning of this assertion we will need to study the concept of *countability* which we now define.

DEFINITION. A set S is said to be *countable* if its elements can be put into $1:1$ correspondence with an infinite subset of the set P of positive integers.

To prove that a set S is countable we need only show that there *exists* a $1:1$ correspondence between the elements of S and the elements of some infinite subset of P (possibly P itself). We are not obliged to state the rule of correspondence, although we often are able to do so by listing the members of S in sequence form, i.e., in such a way that each member of S is certainly in the list and corresponds to the positive integer which identifies its location in the sequence. As an example, we have the set S of *odd* positive integers:

$$S = \{1, 3, 5, 7, 9, \ldots\}.$$

Even though we see before our eyes only a few elements of the set S, we can definitely say what position in the sequence is occupied by any odd positive integer. Looking at the correspondence

$$S = \{1, 3, 5, 7, 9, \ldots\}$$
$$\updownarrow \ \updownarrow \ \updownarrow \ \updownarrow \ \updownarrow$$
$$P = \{1, 2, 3, 4, 5, \ldots\},$$

we can readily see that each element in S is 1 less than twice the positive integer to which it corresponds, i.e., the positive integer which gives its position in the sequence. Thus, for example, the odd positive integer 101 corresponds to a positive integer n, where $101 = 2n - 1$ and $n = 51$; 101 is the 51st term in the sequence of odd positive integers.

Conversely, if we wish to know, say, the 32nd element of S, we immediately see that it is $2 \cdot 32 - 1 = 63$. The amazing thing about this example is that it shows us that the set P of positive integers can be put into 1 : 1 correspondence with one of its subsets; in short, there are "as many" odd positive integers as there are positive integers.

We shall now show that the rational numbers are countable, but first we shall show that the set of positive rationals is countable by considering the following array:

$$1/1 \quad 1/2 \quad 1/3 \quad 1/4\ldots$$

$$2/1 \quad 2/2 \quad 2/3 \quad 2/4\ldots$$

$$3/1 \quad 3/2 \quad 3/3 \quad 3/4\ldots$$

$$4/1 \quad 4/2 \quad 4/3 \quad 4/4\ldots$$
$$\vdots \qquad \vdots \qquad \vdots \qquad \vdots$$

which has infinitely many rows and columns. First we note that every positive rational appears in this array; for example, 143/85 appears in the 143rd row and 85th column and, in general, m/n appears in the mth row and nth column. The arrows indicate a scheme for arranging the positive rationals in a sequence in such a way that the position number of each positive rational will be definitely known. The 1 : 1 correspondence may be written as follows:

$$\frac{1}{1} \quad \frac{1}{2} \quad \frac{2}{1} \quad \frac{3}{1} \quad \frac{2}{2} \quad \frac{1}{3} \quad \frac{1}{4} \quad \frac{2}{3} \quad \frac{3}{2}\cdots$$
$$\updownarrow \quad \updownarrow \quad \updownarrow \quad \updownarrow \quad \updownarrow \quad \updownarrow \quad \updownarrow \quad \updownarrow \quad \updownarrow$$
$$1 \quad 2 \quad 3 \quad 4 \quad 5 \quad 6 \quad 7 \quad 8 \quad 9\ldots$$

We note the following properties of this array:

(i) The first diagonal contains 1 element, the second diagonal contains 2 elements, the third diagonal contains 3 elements, and so on.

(ii) The total number of elements in the first k diagonals is $1 + 2 + 3 + \cdots + k = k(k + 1)/2$. (Prove this! Use the axiom of induction.)

(iii) Each fraction appearing on the *first* diagonal has 2 as the sum of its numerator and denominator, each fraction appearing on the *second* diagonal has 3 as the sum of its numerator and denominator and, in general, each fraction appearing on the kth diagonal has $k + 1$ as the sum of its numerator and denominator.

(iv) The arrows point downward on any *even*-numbered diagonal and the sum of the numerator and denominator of any fraction on an *even*-numbered diagonal is *odd*. The reverse holds true for *odd*-numbered diagonals.

To *verify* that the set of positive rationals is countable let us take a particular rational number, say 27/38, and see if we can locate it in the sequence 1/1, 1/2, 2/1, 3/1, 2/2, 1/3, . . . without writing out the sequence. Since $27 + 38 = 65$, 27/38 appears on the 64th diagonal, and since 64 is even, the arrows point *downward* on this diagonal, starting with 1/64 in the first row. Then, starting with 1/64 and following the arrows through 2/63, 3/62, 4/61, . . . , we find that 27/38 is the 27th element down the 64th diagonal. Since the first 63 diagonals will have been traversed before starting down the 64th, we calculate

$$\frac{63 \cdot (63 + 1)}{2} + 27 = 2043,$$

and conclude that 27/38 is the 2043rd rational number in the sequence 1/1, 1/2, 2/1, 3/1, . . . ; in other words, 27/38 corresponds to 2043. To find the positive integer to which 31/15 corresponds we observe that $31 + 15 = 46$, so that 31/15 appears on the 45th diagonal, and since 45 is odd, the arrows point upward on this diagonal, starting with 45/1 in the first column. Then, starting with 45/1 and following the arrows through 44/2, 43/3, 42/4, . . . , we find that 31/15 is the 15th element up the 45th diagonal. Since the first 44 diagonals will have been traversed before starting up the 45th, we can calculate

$$\frac{44 \cdot (44 + 1)}{2} + 15 = 1005,$$

and say that 31/15 is the 1005th rational number in the sequence 1/1, 1/2, 2/1, 3/1, . . . ; in other words, 31/15 corresponds to 1005. It should now be clear that for any positive rational it is possible to find the positive integer to which it corresponds.

It is also possible to find the rational number corresponding to any given positive integer. For example, to find the rational number corresponding to the positive integer 552 we would first find the largest positive integer k such that

$$\frac{k \cdot (k + 1)}{2} \leq 552.$$

By trial it turns out that $k = 32$, for

$$\frac{32 \cdot (32 + 1)}{2} = 528,$$

while

$$\frac{33 \cdot (33 + 1)}{2} = 561.$$

Thus the 552nd rational number in the sequence must appear on the 33rd diagonal, a diagonal on which each fraction has a numerator and denominator with a sum of 34 and on which the arrows point upward, starting with 33/1. Thus, since $552 - 528 = 24$, the 552nd rational number in the sequence must be the 24th fraction up the 33rd diagonal, namely, 10/24. It has now been *verified* that the set of positive rationals (including all duplications) is countable. Clearly, a rule can be devised for putting the set of *all* rationals (including duplications) into 1 : 1 correspondence with the positive integers. Such a rule might be

$$0 \quad \frac{1}{1} \quad -\frac{1}{1} \quad \frac{1}{2} \quad -\frac{1}{2} \quad \frac{2}{1} \quad -\frac{2}{1} \quad \frac{3}{1} \quad -\frac{3}{1} \ldots$$
$$\updownarrow \quad \updownarrow \quad \updownarrow \quad \updownarrow \quad \updownarrow \quad \updownarrow \quad \updownarrow \quad \updownarrow \quad \updownarrow$$
$$1 \quad 2 \quad 3 \quad 4 \quad 5 \quad 6 \quad 7 \quad 8 \quad 9$$

A positive rational formerly corresponding to a positive integer n now corresponds to $2n$ and its negative corresponds to $2n + 1$. Certainly, then, a rule of correspondence is known; according to this rule every rational corresponds to one and only one positive integer and every positive integer corresponds to one and only one rational. The set of rational numbers, including duplications, is therefore countable. It should be clear that if we delete all duplications in the sequence

$$0, \frac{1}{1}, \frac{-1}{1}, \frac{1}{2}, \frac{-1}{2}, \frac{2}{1}, \frac{-2}{1}, \ldots$$

and at the same time delete from the sequence 1, 2, 3, 4, . . . all positive integers to which they correspond, the subsequence that remains will contain *all* rationals, without duplication, and the terms of the subsequence will be in 1 : 1 correspondence with the infinite subset of positive integers that remain. Thus the set of all rationals is countable, as we wished to show.

We shall now give some indication as to the relative sizes of the set of rationals and the set of irrational numbers. Let us consider the set of all real numbers between 0 and 1 and let us *assume that this set is countable.* Then, if our assumption is correct, it is theoretically possible to list these reals as a sequence in which each one of them occupies a definite position in the sequence:

$$1 \leftrightarrow .a_{11}a_{12}a_{13}a_{14}a_{15} \ldots$$
$$2 \leftrightarrow .a_{21}a_{22}a_{23}a_{24}a_{25} \ldots$$
$$3 \leftrightarrow .a_{31}a_{32}a_{33}a_{34}a_{35} \ldots$$
$$4 \leftrightarrow .a_{41}a_{42}a_{43}a_{44}a_{45} \ldots$$
$$5 \leftrightarrow .a_{51}a_{52}a_{53}a_{54}a_{55} \ldots$$
$$\vdots \quad \vdots \quad \vdots \quad \vdots \quad \vdots$$

This sequence is a sequence of infinite decimals between 0 and 1, where a_{ij} denotes the jth digit of the ith decimal. Remember, this sequence is assumed to contain *all* the real numbers between 0 and 1. Let us now *construct* an infinite decimal $d = .c_1c_2c_3c_4c_5 \ldots$, where $c_1 \neq a_{11}, c_1 \neq 0$, and $c_1 \neq 9$; $c_2 \neq a_{22}$, $c_2 \neq 0$, and $c_2 \neq 9$; $c_3 \neq a_{33}$, $c_3 \neq 0$, and $c_3 \neq 9$; and so on. There are at least 7 and at most 8 ways to select each c_i. The infinite decimal d is certainly a real number between 0 and 1, but it is also certainly not in the sequence, for it differs from each infinite decimal in the sequence in at least one digit. Since, if our assumption were true, d would have to occupy a definite place in the sequence, our assumption is false and the set of real numbers between 0 and 1 is *not* countable. Thus it is readily seen that the set of *all* real numbers is also not countable.

Now let R, S, and T represent the set of all rational numbers, the set of all irrational numbers, and the set of all real numbers, respectively. Set R is countable, as we have shown, so that we may list its members in sequence form:

$$R = \{r_1, r_2, r_3, r_4, \ldots\}.$$

Assume that S is also countable so that we may write

$$S = \{s_1, s_2, s_3, s_4, \ldots\}.$$

Therefore, if our assumption is correct, the set $R \cup S$ is countable, for we may set up the 1 : 1 correspondence

$$
\begin{array}{cccccc}
r_1 & s_1 & r_2 & s_2 & r_3 & s_3 \ldots \\
\updownarrow & \updownarrow & \updownarrow & \updownarrow & \updownarrow & \updownarrow \\
1 & 2 & 3 & 4 & 5 & 6
\end{array}
$$

However, $R \cup S = T$, the set of all reals, and since T is known to be not countable, our assumption that S is countable must be false. We now have a comparison between the size of R and the size of S; R is countable and S is not. But, if the size of R is infinite, what is the size of S; what is larger than infinity? It may be mentioned that there are many different infinities and the interested reader may learn about them by studying a very fascinating kind of arithmetic called *transfinite* arithmetic, whose numbers are these infinities.*

9–8 Completeness of the reals. It has been mentioned (Section 8–8) that the rational number system is not complete; that is, there are holes in it. We earlier announced our intention of constructing a new number system, the real number system, which is complete. We now must define what is meant when it is said that the real number system is complete and we must try to convince the reader that the real number system is in fact complete. At present we can say only that we have filled in holes between the repeating decimals with nonrepeating decimals, but we would only be conjecturing if we said that no holes remain to be filled.

We shall not try to *prove* that all the holes have been filled, but shall merely assert that they have, indeed, been filled and try to indicate how this fact might be proved.

Consider first how the real number system was constructed:

(i) We started with the rationals as our basic building blocks and considered the set C of all C-sequences of rational numbers.

(ii) We defined an equivalence relation $\mathcal{R}$ on the set C, where $x\mathcal{R}y$ implied that $x - y$ was a null sequence, and conversely.

(iii) We partitioned the set C by using the equivalence relation $\mathcal{R}$. We denoted the family of equivalence sets by $\mathcal{C}$ and called this family the *reals*.

(iv) We defined operations $+$ and $\cdot$ on $\mathcal{C}$ and showed that $\{\mathcal{C}; +, \cdot\}$ is a number system and also a field.

(v) We showed that there is a $1 : 1$ correspondence between the set of all infinite decimals and the reals which enabled us to define operations on the infinite decimals in such a way as to make the $1 : 1$ correspondence an isomorphism with respect to each operation. We then elected to use the infinite decimals as symbols for the real numbers.

What we did, then, amounts to this: We started with the rationals and constructed a larger set of numbers (the reals) which has the rationals as a subset. Mathematicians are in the habit of calling this process "extending the rationals to the reals." You may now ask: "If we started with the

---

* J. Houston Banks, *Elements of Mathematics*, New York, Allyn and Bacon, 1956 (Sec. 7.7).

William L. Schaaf, *Basic Concepts of Elementary Mathematics*, New York, Wiley, 1960 (Ch. 5, pp. 148–154).

reals as our basic building blocks, could we, by applying a similar sequence of steps, extend the reals to a larger set of numbers which have the reals as a subset?" The answer is that no larger set having the reals as a subset can be obtained by reapplying the process that was used in extending the rationals to the reals; this is what we mean when we say that the real number system is "complete."

INDEX